# NONLINEAR DYNAMICS, MATHEMATICAL BIOLOGY, AND SOCIAL SCIENCE

# NONLINEAR DYNAMICS, MATHEMATICAL BIOLOGY, AND SOCIAL SCIENCE

**Joshua M. Epstein**
Senior Fellow, Economic Studies Program,
The Brookings Institution, and
Member, External Faculty, Santa Fe Institute

Lecture Notes Volume IV

Santa Fe Institute
Studies in the Sciences of Complexity

*The Advanced Book Program*

CRC Press
Taylor & Francis Group
Boca Raton London New York

CRC Press is an imprint of the
Taylor & Francis Group, an **informa** business
A CHAPMAN & HALL BOOK

Director of Publications, Santa Fe Institute: Ronda K. Butler-Villa
Production Manager, Santa Fe Institute: Della L. Ulibarri
Publication Assistant, Santa Fe Institute: Marylee Thomson

First published 1997 by Addison-Wesley Publishing Company

Published 2019 by CRC Press
Taylor & Francis Group
6000 Broken Sound Parkway NW, Suite 300
Boca Raton, FL 33487-2742

First issued in hardback 2019

*CRC Press is an imprint of the Taylor & Francis Group, an informa business*



Visit the Taylor & Francis Web site at
http://www.taylorandfrancis.com

and the CRC Press Web site at
http://www.crcpress.com

ISBN 13: 978-0-367-32001-0 (hbk)
ISBN 13: 978-0-201-41988-7 (pbk)

This volume was typeset using TEXtures on a Macintosh IIsi computer.

## About the Santa Fe Institute

The *Santa Fe Institute* (SFI) is a private, independent, multidisciplinary research and education center, founded in 1986. Since its founding, SFI has devoted itself to creating a new kind of scientific research community, pursuing emerging syntheses in science. Operating as a small, visiting institution, SFI seeks to catalyze new collaborative, multidisciplinary projects that break down the barriers between the traditional disciplines, to spread its ideas and methodologies to other institutions, and to encourage the practical applications of its results.

All titles from the *Santa Fe Institute Studies in the Sciences of Complexity* series will carry this imprint which is based on a Mimbres pottery design (circa A.D. 950–1150), drawn by Betsy Jones. The design was selected because the radiating feathers are evocative of the outreach of the Santa Fe Institute Program to many disciplines and institutions.

Santa Fe Institute
**Studies in the Sciences of Complexity**

**Proceedings Volumes**

| Vol. | Editors | Title |
|---|---|---|
| I | D. Pines | Emerging Syntheses in Science, 1987 |
| II | A. S. Perelson | Theoretical Immunology, Part One, 1988 |
| III | A. S. Perelson | Theoretical Immunology, Part Two, 1988 |
| IV | G. D. Doolen et al. | Lattice Gas Methods for Partial Differential Equations, 1989 |
| V | P. W. Anderson, K. Arrow, & D. Pines | The Economy as an Evolving Complex System, 1988 |
| VI | C. G. Langton | Artificial Life: Proceedings of an Interdisciplinary Workshop on the Synthesis and Simulation of Living Systems, 1988 |
| VII | G. I. Bell & T. G. Marr | Computers and DNA, 1989 |
| VIII | W. H. Zurek | Complexity, Entropy, and the Physics of Information, 1990 |
| IX | A. S. Perelson & S. A. Kauffman | Molecular Evolution on Rugged Landscapes: Proteins, RNA and the Immune System, 1990 |
| X | C. G. Langton et al. | Artificial Life II, 1991 |
| XI | J. A. Hawkins & M. Gell-Mann | The Evolution of Human Languages, 1992 |
| XII | M. Casdagli & S. Eubank | Nonlinear Modeling and Forecasting, 1992 |
| XIII | J. E. Mittenthal & A. B. Baskin | Principles of Organization in Organisms, 1992 |
| XIV | D. Friedman & J. Rust | The Double Auction Market: Institutions, Theories, and Evidence, 1993 |
| XV | A. S. Weigend & N. A. Gershenfeld | Time Series Prediction: Forecasting the Future and Understanding the Past, 1994 |
| XVI | G. Gumerman & M. Gell-Mann | Understanding Complexity in the Prehistoric Southwest, 1994 |
| XVII | C. G. Langton | Artificial Life III, 1994 |
| XVIII | G. Kramer | Auditory Display, 1994 |
| XIX | G. Cowan, D. Pines, & D. Meltzer | Complexity: Metaphors, Models, and Reality, 1994 |
| XX | D. H. Wolpert | The Mathematics of Generalization, 1995 |
| XXI | P. E. Cladis & P. Palffy-Muhoray | Spatio-Temporal Patterns in Nonequilibrium Complex Systems, 1995 |
| XXII | H. Morowitz & J. L. Singer | The Mind, The Brain, and Complex Adaptive Systems, 1995 |
| XXIII | B. Julesz & I. Kovács | Maturational Windows and Adult Cortical Plasticity, 1995 |
| XXIV | J. A. Tainter & B. B. Tainter | Economic Uncertainty and Human Behavior in the Prehistoric Southwest, 1995 |
| XXV | J. Rundle, D. Turcotte, & W. Klein | Reduction and Predictability of Natural Disasters, 1996 |
| XXVI | R. K. Belew & M. Mitchell | Adaptive Individuals in Evolving Populations: Models and Algorithms, 1996 |

**Lectures Volumes**

| Vol. | Editor | Title |
|---|---|---|
| I | D. L. Stein | Lectures in the Sciences of Complexity, 1989 |
| II | E. Jen | 1989 Lectures in Complex Systems, 1990 |
| III | L. Nadel & D. L. Stein | 1990 Lectures in Complex Systems, 1991 |
| IV | L. Nadel & D. L. Stein | 1991 Lectures in Complex Systems, 1992 |
| V | L. Nadel & D. L. Stein | 1992 Lectures in Complex Systems, 1993 |
| VI | L. Nadel & D. L. Stein | 1993 Lectures in Complex Systems, 1995 |

**Lecture Notes Volumes**

| Vol. | Author | Title |
|---|---|---|
| I | J. Hertz, A. Krogh, & R. Palmer | Introduction to the Theory of Neural Computation, 1990 |
| II | G. Weisbuch | Complex Systems Dynamics, 1990 |
| III | W. D. Stein & F. J. Varela | Thinking About Biology, 1993 |
| IV | J. M. Epstein | Nonlinear Dynamics, Mathematical Biology, and Social Science, 1997 |
| V | H. F. Nijhout, L. Nadel & D. L. Stein | Pattern Formation in the Physical and Biological Sciences, 1997 |

**Reference Volumes**

| Vol. | Author | Title |
|---|---|---|
| I | A. Wuensche & M. Lesser | The Global Dynamics of Cellular Automata: Attraction Fields of One-Dimensional Cellular Automata, 1992 |

*Dedicated in loving memory to my father*

*Joseph Epstein*

*(1917-1993)*

# Contents

# Introduction

This book is based on a series of lectures I gave at the 1992 Santa Fe Institute Complex Systems Summer School, and on my Princeton University "Complex Systems, Simple Models" course, offered in academic years 1991–92 through 1993–94. A goal of my teaching, and of this book, is to impart the mathematical tools and, as important, the *impulse* to build simple models of complex processes falling outside the artificial confines of the established fields. Many fascinating and important problems cry out for rigorous interdisciplinary study. And recent advances in scientific computing have made the construction and "experimental" study of dynamical systems remarkably easy. The stage, in short, is set for new synthetic work, indeed for a new discipline or, perhaps, transdiscipline.

Three themes run through these lectures. The first is that simple models can illuminate essential dynamics of complex, and crucially important, social systems. The second is that mathematical biology offers a powerful, and hitherto underexploited, perspective on both interstate and intrastate social dynamics. The third theme is the unifying power of mathematics, and specifically, of nonlinear dynamical systems theory; formal analogies between seemingly disparate social and biological phenomena are highlighted. One overarching aim is to help stimulate something of a reconstruction in mathematical social science, relaxing—in some cases abandoning—the predominant assumption of perfectly informed utility maximization, and exploring social dynamics from such perspectives as epidemiology and ecosystem science.

## OVERVIEW OF THE LECTURES

There are six lectures. The first is entitled, "On the Mathematical Biology of Arms Races, Wars, and Revolutions." Here, some of the book's recurrent themes are first sounded. It is demonstrated, I believe for the first time, that the most famous equations in the mathematical theories of war (Lanchester's equations) and arms races (Richardson's equations) are both specializations of the famous Lotka-Volterra ecosystem model. The essay introduces the related idea that explosive processes of civil violence—revolutions—might be modeled as epidemics, using yet other parametrizations of Lotka-Volterra. To me, it is surprising and interesting that the Lotka-Volterra ecosystem equations have *anything* to say about war, arms races, or revolutions. Of course, to claim that these simple equations say *everything* on such complex topics would be foolish. And, in subsequent lectures, I move beyond them.

Lecture 2 delves further into the mathematical theory of combat. Drawing on mathematical biology and, of course, history, this lecture offers what I believe to be a fundamental critique of the dominant approach, based on the equations of F. W. Lanchester. Lanchester Theory, by which I mean the original equations and their contemporary extensions, produces anomalous results, mathematically precluding important observed behaviors, like the trading of space for time. These deep problems arise because the belligerents, as idealized in the theory, are completely non-adaptive. War is, I argue, precisely a process of coadaptation, though the contestants bear a much closer resemblance to Ashby's cat than to *Homo economicus*. My own Adaptive Dynamic Model tries to capture this as simply as possible, in the process overcoming the anomalies of Lanchester Theory.

The "nonlinear dynamics of hope" is illustrated in lecture 3. By way of introduction, many applications of nonlinear dynamics show that complex systems can be poised at the brink of disaster; small perturbations in crucial variables can produce cascade extinctions in ecosystems, devastating epidemics in human populations, or ozone holes in the atmosphere, all unhappy events. Sensitive dependence is bad news. Perhaps it is salient that we call the area "catastrophe theory" and not, for instance, "miracle theory." Well, this lecture invites us to consider the flipside of the nonlinear coin: can we identify cases in which the right, small local perturbation can produce counterintuitive explosions of happy events? I think so.

As one example, there is an idea called "collective security" (CS) that is receiving wide attention. Imagine three countries, A, B, and C. *Perfect* CS would then operate as follows: If A attacks B, C allocates all force to B; if B attacks C,

A allocates all force to C; and so on. The general rule is simply that *the odd man out instantly allocates all force to the attacked party.*

Now, many academic political scientists and statesmen dismiss *any* form of collective security because this *perfect* form is implausibly altruistic. But, as nonlinear dynamicists, we ask: What about a tiny bit of collective security, a highly diluted form of altruism? The lecture shows that in arms race models sufficiently nonlinear to produce really volatile dynamics, highly diluted, or imperfect, collective security regimes can damp the explosive oscillations and induce convergence to stable equilibria below initial armament levels. Put differently, the injection of *tiny* degrees of altruism can profoundly calm the otherwise volatile dynamics. The benefits of the system are very great and, because of the nonlinearity, the level of risk to individual participants is very low. Here, sensitive dependence is good news!

Whether the next lecture bears good news or bad depends, I suppose, on one's political leanings. It examines the analogy between epidemics (for which a well-developed mathematical theory exists) and processes of explosive social change, such as revolutions (for which no comparable body of mathematical theory exists). Are revolutions "like" epidemics? If one thinks of the revolutionary idea as the infection, the revolutionaries as the infectives, the public health authorities as the power elite, and social indoctrination as inoculation, then an analogy begins to take shape. It is developed in the fourth lecture, with, I hope, some novel political interpretations. The analogy to epidemics, which are *nonlinear* threshold processes, may help explain how small changes in political conditions—marginal diminutions in central authority—can catalyze explosive social transformations, much to the surprise of elites and revolutionaries alike! The model also suggests the existence of social bifurcation points at which repression abruptly changes from being stabilizing to being destabilizing and inflaming revolutionary sentiment.

The fifth lecture combines arms race and epidemiology perspectives in building a simple model of the spread of drug addiction in an idealized community, revealing basic, and perhaps counterintuitive, relationships between legalization, prices, and crime. The analysis suggests once again the relevance to social science of seemingly remote fields like mathematical epidemiology and ecosystem science, and it tries to illustrate how simple models are built, and explored using methods of nonlinear analysis.

These methods are the topic of lecture 6. Entitled "An Introduction to Nonlinear Dynamical Systems," it consists entirely of mathematics. The aim is to offer a concentrated course in the qualitative theory of nonlinear autonomous differential systems, beginning with linearized stability analysis and moving efficiently through Lyapunov functions, limit cycles, the Poincaré-Bendixson and Hopf Bifurcation Theorems, Poincaré maps, various negative tests, and on to Index Theory and the celebrated Poincaré-Hopf Theorem from differential topology. I see this as a coherent body of mathematics, much of which is quite beautiful, and powerful when applied to social dynamics.

I hasten to point out that lecture 6 is not designed as a mathematical foundation for the other lectures. Not all the techniques developed there are applied in other lectures. Index Theory, for example, is developed for the sheer joy of it, not because I use it elsewhere, though its surprising applications to mathematical ecology and economics are noted. In turn, not all techniques used in other lectures are covered in lecture 6. Lectures 2 and 3, for instance, involve difference—rather than differential—equations, which are not treated in lecture 6. Lecture 6, then, is a free-standing essay offering a particular development of nonlinear dynamical systems, from linearized stability analysis through the Poincaré-Hopf Index Theorem via results of longstanding mathematical interest, such as Hilbert's 16th Problem and Brouwer's Fixed Point Theorem.[1]

One final point regarding the lectures should be made. While I often use history or empirical studies to argue for the qualitative plausibility of a model, no new data bases are assembled or statistical tests performed. As often occurs in science, theory may ultimately inspire the collection of data and the performance of tests. But these lectures are purely theoretical, the goal being to demonstrate to social, physical, and natural scientists that simple mathematical models can provide *insight* into a wide range of complex social processes and that mathematical biology and nonlinear dynamical systems theory in particular offer the social theorist powerful conceptual and analytic tools.[2]

## THE LARGER INTELLECTUAL LANDSCAPE

These, of course, are not the only methods available for the study of social phenomena. And, in my Princeton course, I gave equal time to the agent-based modeling techniques employed in *Growing Artificial Societies: Social Science From the Bottom Up*, co-authored by myself and Robert Axtell. Both nonlinear dynamical systems and agent-based models deserve a place in any "complexity curriculum." But the former techniques are the ones employed here. For agent-based models of social systems, see Epstein and Axtell (1996) and the growing literature cited there.

## ACKNOWLEDGMENTS

A number of colleagues and institutions deserve special thanks. For careful reviews of the manuscript or portions thereof, for stimulating discussions, encouragement, or

[1]The lecture assumes familiarity with vector calculus, linear differential equations, eigenvalue-eigenvector methods, phase plane analysis, and certain elements of complex variables, real analysis, and partial differential equations. Portions of the other lectures also assume exposure to certain of these topics.

[2]On insight, as against prediction, as a goal of modeling, see Hirsch (1984).

advise, I thank Robert Axelrod, Robert Axtell, Bruce G. Blair, John Casti, Malcolm DeBevoise, George Downs, Samuel David Epstein, Marcus W. Feldman, Duncan Foley, Murray Gell-Mann, Atlee Jackson, Jean-Pierre Langlois, Steven McCarroll, Elaine C. McNulty, Gottfried Mayer-Kress, Benoit Morel, Lee Segel, Carl Simon, Daniel Stein, Arthur S. Wightman, and H. Peyton Young. I thank the Princeton University Council on Science and Technology for funding, and the Woodrow Wilson School for hosting, my course. I am grateful to the Brookings Institution for its support and especially to John D. Steinbruner for the climate of unfettered inquiry in which this research was conducted. I thank Daniel Stein for organizing the Santa Fe Institute Complex Systems Summer School. I offer deep thanks also to my former SFI and Princeton students. For expert assistance in preparing the manuscript, I thank Trisha Brandon. I am grateful to Ronda K. Butler-Villa for editing the manuscript, and to Della L. Ulibarri for production assistance.

Finally, for their love and support, I thank my wife Melissa, our daughter Anna Matilda, my mother Lucy, and my brother Sam.

The views expressed in this book are those of the author and should not be ascribed to the persons or organizations acknowledged above.

# LECTURE 1
# On The Mathematical Biology of Arms Races, Wars, and Revolutions

In this opening lecture, I will attempt a unifying overview of certain social phenomena—war, arms racing, and revolution—from the perspective of mathematical biology, a field which, in my view, must ultimately subsume the social sciences.[3] Unfortunately, few social scientists are exposed to mathematical biology, specifically the dynamical systems perspective pioneered by Alfred Lotka, Vito Volterra, and others. In turn, few mathematical biologists have considered the application of mathematical biology to problems of human society.[4]

Particularly in areas of interstate and intrastate conflict is there a need to explore formal analogies to biological systems. On the topic of animal behavior and human warfare, the anthropologist Richard Wrangham observes,

[3]The perspective taken here, however, is quite distinct from that taken by Edward O. Wilson, in his book *Sociobiology* (1980). Specifically, I do not discuss the role of genes in the control of human social behavior. Rather, the argument is that macro social behaviors such as war, revolution, arms races, and the spread of drugs may conform well to equations of mathematical biology—ecology and epidemiology in particular. Perhaps "socioecology" would be a suitable name for this level of analysis.

[4]For a notable exception, see Cavalli-Sforza and Feldman (1981). See also the innovative and understudied works, Rashevsky (1947) and Rashevsky (1949).

> "The social organization of thousands of animals is now known in considerable detail. Most animals live in open groups with fluid membership. Nevertheless there are hundreds of mammals and birds that form semi-closed groups, and in which long-term intergroup relationships are therefore found. These intergroup relationships are known well. In general they vary from benignly tolerant to intensely competitive at territorial borders. The striking and remarkable discovery of the last decade is that only two species other than humans have been found in which breeding males exhibit systematic stalking, raiding, wounding and killing of members of neighboring groups. They are the chimpanzee (*Pan troglodytes*) and the gorilla (*Pan gorilla beringei*) (Wrangham, 1985). In both species a group may have periods of extended hostility with a particular neighboring group and, in the only two long-term studies of chimpanzees, attacks by dominant against subordinate communities appeared responsible for the extinction of the latter.
>
> "Chimpanzees and gorillas are the species most closely related to humans, so close that it is still unclear which of the three species diverged earliest (Ciochon & Chiarelli, 1983). The fact that these three species share a pattern of intergroup aggression that is otherwise unknown speaks clearly for the importance of a biological component in human warfare" (Wrangham, 1988, p.78).

Although man has engaged in arms racing, warring, and other forms of organized violence for all of recorded history, we have comparatively little in the way of formal theory. Mathematical biology may provide guidance in developing such a theory. Wrangham writes, "Given that biology is in the process of developing a unified theory of animal behavior, that human behavior in general can be expected to be understood better as a result of biological theories, and that two of our closest evolutionary relatives show human patterns of intergroup aggression, there is a strong case for attempting to bring biology into the analysis of warfare. At present, there are few efforts in this direction."[5] I would like to see more effort, specifically more mathematical effort, in this direction and hope to stimulate some interest among you. To convince you that there might conceivably be some "unified field theory" worth pursuing, I want to share some observations with you. To set them up, a little background is required.

The fundamental equations in the mathematical theory of arms races are the so-called Richardson equations, named for the British applied mathematician and social scientist Lewis Frye Richardson, who first published them in 1939.[6] The fundamental equations in the mathematical theory of combat (warfare itself, as against peacetime arms racing) were published in 1916 by Frederick William Lanchester.[7]

[5] Wrangham (1988, p.78).

[6] Richardson (1939) and (1960).

[7] See Lanchester (1916). For a contemporary discussion with references, see Epstein (1985).

The formal theory of interstate conflict, to the extent there is one, rests on these twin pillars, if you will. Meanwhile, the classic equations of mathematical ecology are the Lotka-Volterra equations.

In light of the remarks above, I find the following fact intriguing: The Richardson and Lanchester models of human conflict are, mathematically, specializations of the Lotka-Volterra ecosystem equations.

Before proceeding, I must make one point unmistakably clear. I do not claim that any of these models is really "right" in a physicist's sense. They are illuminating abstractions. I think it was Picasso who said, "Art is a lie that helps us see the truth." So it is with these simple models. They continue to form the conceptual foundations of their respective fields. They are universally taught; mature practioners, knowing full-well the models' approximate nature, nonetheless entrust to them the formation of the student's most basic intuitions. And this because, like idealizations in other sciences—idealizations that are ultimately "wrong"—they efficiently capture qualitative behaviors of overarching interest. That these ecosystem and, say, arms race equations should look at all alike is unexpected. That, on closer inspection, they are virtually identical is, to me, really quite interesting. Let me go a bit further.

Under yet other parameter settings, the Lotka-Volterra equations yield standard models of epidemics. And, in other lectures, I will argue that social revolutions and illicit drugs may well spread in a strictly analogous way or—at the very least—that an epidemiological perspective on such social processes is promising. Once more, the point is simply that social science might learn a lot from mathematical biology and, conceivably, might inherit some of its apparent unity.

Let me now introduce the Lotka-Volterra equations and show how the classic arms race and war models fall out as special cases. Then, I will explore the analogy between revolutions and epidemics. In subsequent lectures, we will move beyond these simple—too simple—models.

## THE LOTKA-VOLTERRA WORLD

The Lotka-Volterra equations are as follows:

$$\begin{aligned} \dot{x}_1 &= x_1(r_1 - a_{11}x_1 + a_{12}x_2)\,, \\ \dot{x}_2 &= x_2(r_2 + a_{21}x_1 - a_{22}x_2)\,. \end{aligned} \tag{1.1}$$

In discussing these equations, I will freely invoke nonlinear dynamical systems terminology presented in lecture 6.[8] Turning now to system (1.1), $x_i(t)$ is the species $i$ population at time $t$; the $a$'s and $r$'s are real parameters.

[8] Under the name, "quadratic model," equivalent equations and a number of specializations—including combat variants—are discussed in Beltrami (1987).

If all $a_{ij}$'s equal zero and $r_1, r_2 > 0$, we have unbounded exponential—so-called Malthusian—growth. Since, ultimately, there are limits, for instance, environmental carrying capacities, the terms $a_{11}, a_{22} > 0$ are preceded by a negative sign. Then, in the language of lecture 6, the species are self-inhibiting. Leaving $r_1$ and $r_2$ positive and still assuming $a_{12} = a_{21} = 0$, this assumption yields a logistic approach for each species to the positive phase plane equilibrium

$$(\bar{x}_1, \bar{x}_2) = \left(\frac{r_1}{a_{11}}, \frac{r_2}{a_{22}}\right),$$

a node sink.

Now, life really gets interesting only when species interact, and this involves the cross terms $a_{12}$ and $a_{21}$.

## MUTUALISM

Leaving everything else as is, let us now assume $a_{12}, a_{21} > 0$. In that case our species are said to be in a relationship of mutualism, or reciprocal activation; the population level of one feeds back positively on the growth rate of the other. Bees and flowers—pollinators and pollinatees, if you will—provide examples. There are many others.

Setting $\dot{x}_1 = \dot{x}_2 = 0$, the interior equilibrium conditions are

$$\begin{aligned} r_1 - a_{11}x_1 + a_{12}x_2 &= 0\,, \\ r_2 + a_{21}x_1 - a_{22}x_2 &= 0\,. \end{aligned} \tag{1.2}$$

Of course, these are also the equilibrium conditions for the linear system:

$$\begin{aligned} \dot{x}_1 &= r_1 - a_{11}x_1 + a_{12}x_2\,, \\ \dot{x}_2 &= r_2 + a_{21}x_1 - a_{22}x_2\,. \end{aligned} \tag{1.3}$$

But this is exactly the famous Richardson model of an arms race! The more bees, the more flowers, and vice versa. It's the same in (1.3), but not quite as idyllic. The more weaponry my adversary has, the more I want, and vice versa, up to some economic—or ecological—limit or carrying capacity.

Richardson's basic idea is that a state's arms race behavior depends on three overriding factors: the perceived external threat, the economic burden of military competition, and the magnitude of grievances against the other party. I discuss these at greater length in lecture 3. Suffice it to say here that $r_1, r_2 > 0$ represent fundamental grievances; $a_{12}, a_{21} > 0$ are the reciprocal activation coefficients (the rates at which each arsenal grows in response to the other); and $a_{11}, a_{22}$ are the self-inhibiting, or damping, terms which Richardson identified with economic fatigue.

Mathematical biologists have long asked how mutualistic populations avoid exploding in what Robert May called an "orgy of mutual benefaction."[9] Likewise, we can ask what mechanism damps the upward action-reaction military dynamic represented in the Richardson model. In each case, self-inhibitory effects must somehow dominate reciprocal activation effects if a stable species equilibrium—or military "balance of power"—is to emerge. Stability analysis bears this out.

Clearly, we can write (1.2) in matrix form $r + Ax = 0, x \in \mathcal{R}^2$. The positive (or interior) equilibrium of system (1.1) and the sole equilibrium of (1.3) is therefore given by $\bar{x} = -A^{-1}r$. For each model, the stability of $\bar{x}$ can be evaluated by the methods of lecture 6.

By a simple translation, the Richardson equations (1.3) are globally asymptotically stable at $\bar{x}$ if and only if $\dot{y} = Ay$ is globally asymptotically stable at the origin, where $y = x - \bar{x}$. From lecture 6, we have the well-known stability criterion

$$\text{Tr } A < 0 \text{ and } \text{Det } A > 0 . \tag{1.4}$$

Now, Richardson's economic fatigue *means* $a_{11}, a_{22} > 0$. So, we have

$$\text{Tr } A = -a_{11} - a_{22} < 0 .$$

And we will have Det $A > 0$ precisely when $a_{11}a_{22} > a_{12}a_{21}$, which is to say that inhibition $(a_{11}a_{22})$ outweighs activation $(a_{12}a_{21})$, confirming our intuition.

One can demonstrate[10] that the eigenvalues of the Jacobian of (1.1) at $\bar{x}$ have negative real parts (indeed, are negative reals) when the same condition is met. An isocline analysis is also revealing. We recall that an isocline is a curve—here a line—where one side's rate of growth is zero; clearly, an equilibrium is a point where isoclines intersect. From (1.2), the isoclines are given by:

$$\begin{aligned} \phi_1(x_1) &= \frac{a_{11}}{a_{12}}x_1 - \frac{r_1}{a_{12}} \text{ (the } x_1 - \text{isocline)} , \\ \phi_2(x_1) &= \frac{a_{21}}{a_{22}}x_1 + \frac{r_2}{a_{22}} \text{ (the } x_2 - \text{isocline)} . \end{aligned} \tag{1.5}$$

For local stability of the equilibrium $\bar{x}$, we require the configuration of figure 1.1. But, this occurs only if the slope of $\phi_1$ exceeds the slope of $\phi_2$, which is to say $a_{11}/a_{12} > a_{21}/a_{22}$, or

$$a_{11}a_{22} > a_{21}a_{12} .$$

Our intuition is again confirmed: stability requires self-inhibition to exceed reciprocal activation in this sense.

[9] May (1981).

[10] Goh (1979).

FIGURE 1.1 Mutualistic Stability

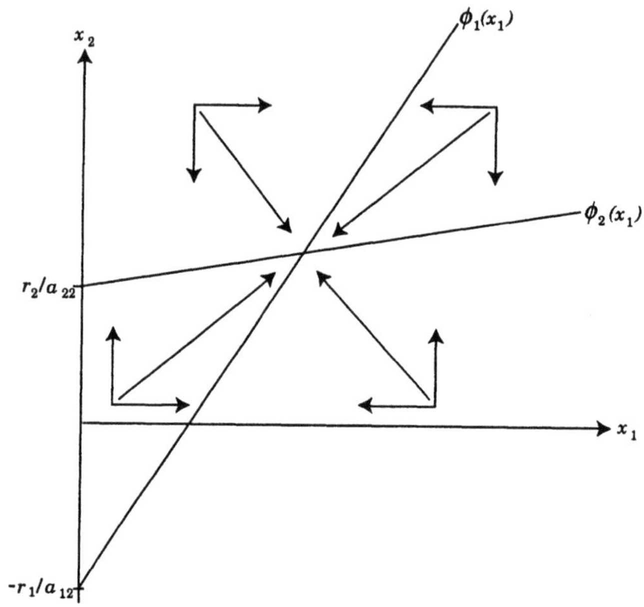


The main point, however, is that the classic Lotka-Volterra model of mutualistic species interaction embeds, in its equilibrium behavior, the classic Richardson arms race model.

## AN ASIDE ON COEVOLUTION

In the models above, of course, the "phenotypes" do not change. In fact, ecosystem dynamics select against certain phenotypes. Roughly speaking, phenotypic frequencies and population levels have interdependent trajectories. This is very clear, for example, in immunology, where antigens and antibodies coevolve in a so-called "biological arms race." But, of course, real arms races work this way, too. Ballistic missiles beget antiballistic missile defenses, which beget various evasion and defense suppression technologies. The machine gun makes cavalry obsolete, giving rise to the "iron horse"—the tank—which begets antitank weapons, which beget special armor, and so on. Michael Robinson's analogy between moth-bat coevolution and the coevolution of World War II air war tactics is apposite.

"Moths and their predators are in an arms race that started millions of years before the Wright brothers made the Dresden raids possible. Butterflies exploit the day, but their 'sisters' the moths dominate the insects' share of the night skies. Few vertebrates conquered night flying. Only a small fraction of bird species, mostly owls and goatsuckers, made the transition. Bats, of course, made it their realm. Many species of bats are skilled 'moth-ers': they pursue them at speed after detecting them with their highly attuned echolocation system. Some moths, however, have developed 'ears' capable of detecting the bat's ultrasonic cries. When they hear a bat coming, the moths take evasive action, including dropping below the bat's track. The parallels of the response of Allied bombers to the radar used by the Germans in World War II are interesting. If we visualize the bombers as the moths, and radars on the ground and in the night-fighter aircraft as bats (a reversal of sizes), the situation is similar. Bombers used rearward-listening radar to detect enemy night fighters. When they detected a fighter, they took evasive action. But heavy bombers, heavily laden, were not very maneuverable. They couldn't dodge about quite as well as moths. Some pilots tried to drop their aircraft into a precipitous dive. Moths also do this; it is easy for them to fold their wings and drop. The next stage in the night-battle escalation is predictable. The night fighter's radar was eventually tuned to detect the bomber's fighter-detector, and thus the bomber itself. Bats have not yet tuned in on moths' ears.

"Bombers also used technological disruption. Night fighters came to be guided to bombers by long-distance radars on the ground. The fighters started winning. But nothing remains static. The ground radars could be jammed by various kinds of radio noise. The technological battle swung the other way. Then the fighters acquired radar. Much like a bat, a fighter emitted and listened to radar signals of its own. These, too, proved to be susceptible to countermeasures, however. The RAF could jam the fighters' radar or 'clutter' it with strips of aluminum foil. Each bomber in a formation dropped one thousand-strip bundle per minute, so that huge clouds of foil foiled the radar. Amazingly, there may be a similar counter-weapon among moths. Some moths can produce ultrasonic sounds that fall within the bats' audio frequency. The moths' voice boxes are paired, one on each side of the thorax; double voices must be particularly confusing. Alien sounds in their waveband could confound the bats, exactly in the same way the foil confounded the fighters.

"The next steps in the bat-versus-moth war may simply be awaiting discovery by some bright researcher; after all, we did not know a lot about echolocation in bats until after World War II. My guess would be that the detector will get more complex to meet the defenses. This may already have happened; bats specializing in moths with ears may have moved to a higher frequency sound outside the moths' hearing range!" (Robinson, 1992, pp. 77–79).

Quite clearly, *levels* of armament (in the international system) and *levels* of population (in an ecosystem) interact, as in the Lotka-Volterra and Richardson models, but *phenotypes* themselves are also changing. In biology, there is a mathematical theory of coevolution.[11] In social science, there isn't. There probably could be, so I simply mention it as a promising direction.

Now, let us shift gears from the mutualistic/arms race variant of (1.1). Specifically, instead of assuming that $a_{12}$ and $a_{21}$ are positive, assume that they are negative.

## COMPETITION

Rearranging slightly, the equations (1.1) take the form

$$\begin{aligned} \dot{x}_1 &= a_{12}x_1x_2 + r_1x_1\left(1 - \frac{x_1}{k_1}\right), \\ \dot{x}_2 &= a_{21}x_1x_2 + r_2x_2\left(1 - \frac{x_2}{k_2}\right), \end{aligned} \tag{1.6}$$

where $k_i \equiv (r_i/a_{ii}) > 0$ is the carrying capacity of the environment for each species. These equations were published in 1934 by the great Russian mathematical biologist G. F. Gause in his book *The Struggle for Existence*. Indeed, he termed $a_{12}$ and $a_{21}$ "coefficients of the struggle for existence."[12]

Now, examining (1.6), each species would exhibit logistic growth to its respective carrying capacity but for these interaction—struggle—terms. Including them, (1.6) gives a picture of uniform mixing of the populations $x_1$ and $x_2$, with contacts proportional to the product $x_1x_2$. Now, however, since the interaction coefficients are negative, each contact *kills* species 1 at rate $a_{12}$ and species 2 at rate $a_{21}$. Quite clearly, a parallel to combat is suggested. But more is true.

In fact, unbeknownst to Gause, (1.6) is an exact form of the famous—and to this day ubiquitous—Lanchester model of warfare![13]

The transition from arms race to war, then, might be seen as a transition from the case of $a_{12}, a_{21} > 0$ to the case of $a_{12}, a_{21} < 0$. In the latter context, the well-known biological "principal of competitive exclusion" simply maps to the military principle that, usually, one side wins and the other side loses. Both these competitive

[11] Roughgarden (1979).

[12] Gause (1934, p. 47).

[13] Lanchester (1916).

exclusion behaviors reflect the mathematical fact that the interior ($x_1, x_2 > 0$) equilibrium of (1.6) is a saddle. The stable equilibrium in the mutualistic—peacetime arms race—case was a node. To the extent these models are correct, then, we can say (*pacem* Poincaré) that war is topologically different from peace; the outbreak of war is a bifurcation from node to saddle.

Thus far we have been exploring a mathematical biology of interstate relations; what about intrastate dynamics? Is there a Lotka-Volterra perspective on revolution, for instance? And, to what biological process might such social dynamics correspond?

## REVOLUTIONS AND EPIDEMICS

Consider the following specialization of (1.1):

$$a_{12} = a_{21} > 0; r_1 = r_2 = a_{11} = a_{22} = 0\,. \tag{1.7}$$

Then (1.1) becomes

$$\begin{aligned}\dot{x}_1 &= -a_{12}x_1x_2\,,\\ \dot{x}_2 &= a_{12}x_1x_2\,,\end{aligned} \tag{1.8}$$

which is the simplest conceivable epidemic model. Now, rather than armament levels, $x_1$ represents the level of susceptibles, and $x_2$ the level of infectives, while the parameter $a_{12}$ is the infection rate, expressing the contagiousness of the infection. Ideal homogeneous mixing, once more, is assumed. If population is constant at $P_0$, then $x_1 = P_0 - x_2$ and we obtain

$$\dot{x}_2 = a_{12}x_2(P_0 - x_2)\,, \tag{1.9}$$

our familiar friend the logistic differential equation. Here, $x_2 = 0$ is an unstable equilibrium; the slightest introduction of infectives, and the disease whips through the whole of society.

A traditional tactic for combating the spread of a disease is removal of infectives. Sometimes, nature does the removing, as with fatal diseases; often, society removes infectives from circulation by quarantine. The simplest possible assumption is that removal is proportional to the size of the infective pool, yielding the following variant of (1.1):

$$\begin{aligned}\dot{x}_1 &= -a_{12}x_1x_2\,,\\ \dot{x}_2 &= a_{12}x_1x_2 - r_2x_2\,,\end{aligned} \tag{1.10}$$

with $r_2 > 0$. This is the famous Kermack-McKendrick (1927) *threshold* epidemic model,[14] so-called because it exhibits the following behavior.

[14] Kermack and McKendrick (1927). For a contemporary statement, see Waltman (1974).

By definition, there is an epidemic outbreak only if $\dot{x}_2 > 0$. But this is to say $a_{12}x_1x_2 - r_2x_2 > 0$, or

$$x_1 > \frac{r_2}{a_{12}} . \tag{1.11}$$

The initial susceptible level $x_1(0)$ must exceed the threshold $\rho \equiv r_2/a_{12}$, sometimes called the relative removal rate, for an epidemic to break out. The fact that epidemics are threshold phenomena has important implications for public health policy and, I will argue below, for social science.

The public health implication, which was very controversial when first discovered, is that *less than* universal vaccination is required to prevent epidemics. By the threshold criterion (1.11), the fraction immunized need only be big enough that the unimmunized fraction—the actual susceptible pool—be below the threshold $\rho$. "Herd immunity," in short, need not require immunization of the entire herd. For instance, diphtheria and scarlet fever require 80-percent immunization to produce herd immunity.[15] Hethcote and Yorke argue that "a vaccine could be very effective in controlling gonorrhea...for a vaccine that gives an average immunity of 6 months, the calculations suggest that random immunization of 1/2 of the general population each year would cause gonorrhea to disappear."[16]

Mathematical epidemic models are discussed more fully in lecture 4. With the above as background, let us now consider the analogy between epidemics (for which a rich mathematical theory exists) and processes of explosive social change, such as revolutions (for which no comparable body of mathematical theory exists). Again, a more careful and deliberate development is given in lecture 4. Here, we simply offer the main idea. It will facilitate exposition to re-label the variables in (1.10). If $S(t)$ and $I(t)$ represent the susceptible and infective pools at time $t$ and if $r$ and $\gamma$ are the infection and removal rates, the basic model is:

$$\begin{aligned} \dot{S} &= -rSI , \\ \dot{I} &= rSI - \gamma I , \end{aligned} \tag{1.12}$$

with epidemic threshold

$$S > \frac{\gamma}{r} = \rho . \tag{1.13}$$

The basic mapping from epidemic to revolutionary dynamics is direct. The infection or disease is, of course, the revolutionary idea. The infectives $I(t)$ are individuals who are actively engaged in articulating the revolutionary vision and in winning over ("infecting") the susceptible class $S(t)$, comprised of those who are receptive to the revolutionary idea but who are not infective (not actively engaged in transmitting the disease to others). Removal is most naturally interpreted as the political imprisonment of infectives by the elite ("the public health authority").

[15] Edelstein-Keshet (1988, p. 255).

[16] Hethcote and Yorke (1980, p. 47).

Many familiar tactics of totalitarian rule can be seen as measures to minimize $r$ (the effective contact rate between infectives and susceptibles) or maximize $\gamma$ (the rate of political removal). Press censorship and other restrictions on free speech reduce $r$, while increases in the rate of domestic spying (to identify infectives) and of imprisonment without trial increase $\gamma$.

Symmetrically, familiar revolutionary tactics—such as the publication of underground literature, or "samizdat"—seek to increase $r$. Similarly, Mao's directive that revolutionaries must "swim like fish in the sea," making themselves indistinguishable (to authorities) from the surrounding susceptible population, is intended to reduce $\gamma$.

## GORBACHEV, DeTOQUEVILLE, AND THE THRESHOLD

Interpreting the threshold relation (1.13), if the number of susceptibles $S_0$ is, in fact, quite close to $\rho$, then even a slight reduction (voluntary or not) in central authority can push society over the epidemic threshold, producing an explosive overthrow of the existing order. To take the example of Gorbachev, the policy of Glasnost obviously produced a sharp increase in $r$, while the relaxation of political repression (e.g., the weakening of the KGB, the release of prominent political prisoners, and the dismantling of Stalin's Gulag system) constituted a reduction in $\gamma$. Combined, these measures evidently depressed $\rho$ to a level below $S_0$, and the "revolutions of 1989" unfolded. Perhaps DeToqueville intuited the threshold relation (1.13), describing this phenomenon, when he remarked that "liberalization is the most difficult of political arts."

As a final element in the analogy, systematic social indoctrination can produce herd immunity to potentially revolutionary ideas. We even see "booster shots" administered at regular intervals—May 1 in Moscow; July 4 in America—on which occasions the order-sustaining myths ("The USSR is a classless workers' paradise"; "Everyone born in America has the same opportunities in life") are ritually celebrated.

Now, as I said before, all these analogies are doubtlessly terribly crude. I certainly do not claim either that any of the models are right or that the dynamical analogies among them are exact. Yet, the very fact that a single ecosystem model—the Lotka-Volterra equations—could specialize to equations that even caricature, however crudely, such basic and important social processes as arms racing, warring, and rebelling is, I believe, very interesting and serves to reinforce the larger point with which I began: social science is ultimately a subfield of biology.

## CONCLUSION

Finally, let me conclude with an admission. I was surprised when I began to notice these connections. But why should we be surprised? In certain non-Western cultures, where our species is seen as "a part *of* nature," where gods—like the sphinx—can be part man and part lion, all these connections between ecosystems and social systems might appear quite unremarkable. But in Western cultures shaped by the Old Testament, where God creates *only* man—not the fishes, birds, and bushes—in *his* own image, man is seen as "apart *from* nature." And, accordingly, we are surprised when our models of fish—or worse yet, of viruses—turn out to be interesting models of man. Perhaps we are true Darwinians more in our heads than in our hearts. Creatures of habit, we are captive to a transmitted and slowly evolving culture. But, of course, this too is "only natural."

# LECTURE 2
# An Adaptive Dynamic Model of Combat

In this lecture I would like to give an introduction to some simple mathematical models of combat, including my own Adaptive Dynamic Model. Here, we are concerned with the *course* of war, rather than the arms races or crises that may precipitate war. Before discussing specifics, it may be well to consider the basic question: What are appropriate goals for a mathematical theory of combat at this point?

First and foremost, we need to be humble. Warfare is complex. Outcomes may depend, perhaps quite sensitively, on technological, behavioral, environmental, and other factors that are very hard to measure before the fact. Exact prediction is really beyond our grasp.

But, that's not so terrible. Theoretical biologists concerned with morphogenesis—the development of pattern—are, in some cases, situated similarly. For the particular leopard, we certainly cannot predict the exact size and distribution of spots. But, certain classes of partial differential equations—reaction-diffusion equations—will generate generic animal coat patterns of the relevant sort. So, we feel that this is the right body of mathematics to be exploring. The same sort of point holds for epidemiologists. Few would claim to be able to predict the exact onset point or severity of an epidemic. Theoreticians seek simple models that will generate a reasonable menu of core qualitative behaviors: threshold eruptions, persistence at endemic levels, recurrence in cycles, perhaps chaotic dynamics. The aim is to produce transparent, parsimonious models that will *generate the core menu of gross qualitative system behaviors.* This, it seems to me, is the sort of claim one would want to make for a mathematical theory of combat.

Now, in classical mechanics, the crucial variables are mass, position, and time. In classical economics, they are price and quantity. War, traditionally, is about territory and, unfortunately, death, or mutual attrition. A respectable model, at the very least, should offer a plausible picture of the relationship between the fundamental processes of attrition and withdrawl (i.e., territorial sacrifice). I will discuss attrition first.

## LANCHESTER'S EQUATIONS

The big pioneer in this general area was Frederick William Lanchester (1868–1945). The eclectic English engineer made contributions to diverse fields, including automotive design and the theory of aerodynamics.[17] He is best remembered for his equations of war, appropriately dubbed the Lanchester equations. First set forth in his 1916 work, *Aircraft in Warfare*, these have a variety of forms, the most renowned of which is called—for reasons that will be given shortly—the Lanchester "square" model.[18] With no air power and no reinforcements, the Lanchester square equations are

$$\begin{aligned} \frac{dR}{dt} &= -bB\,, \\ \frac{dB}{dt} &= -rR\,. \end{aligned} \tag{2.1}$$

Here, $B(t)$ and $R(t)$ are the numbers of "Blue" and "Red" combatants—each of which is an idealized fire source—and $b, r > 0$ are their respective firing effectiveness per shot. Qualitatively, these equations say something intuitively very appealing, indeed, seductive: *The attrition rate of each belligerent is proportional to the size of the adversary.* In the phase plane, the origin is obviously the only equilibrium of (2.1) and the Jacobian of (2.1) at $\bar{x}$ is

$$DF(\bar{x}) = \begin{pmatrix} 0 & -b \\ -r & 0 \end{pmatrix}.$$

[17] Lanchester (1956).

[18] See Lanchester (1916). The same model was apparently developed independently by the Russian M. Osipov (1915).

The eigenvalues are clearly $\pm\sqrt{rb}$. Hence, the origin is a saddle, though the positive quadrant is all we care about. The system (2.1) is, of course, soluble exactly. With $B(0) = B_0$ and $R(0) = R_0$,

$$\begin{aligned} R(t) &= \frac{1}{2}\left[\left(R_0 - \sqrt{\frac{b}{r}}B_0\right)e^{\sqrt{rb}\;t} + \left(R_0 + \sqrt{\frac{b}{r}}B_0\right)e^{-\sqrt{rb}\;t}\right], \\ B(t) &= \frac{1}{2}\left[\left(B_0 - \sqrt{\frac{r}{b}}R_0\right)e^{\sqrt{rb}\;t} + \left(B_0 + \sqrt{\frac{r}{b}}R_0\right)e^{-\sqrt{rb}\;t}\right], \end{aligned} \tag{2.2}$$

with various trajectories for $R$ and $B$ over time. Depending on the parameters $(b, r)$ and the initial values $(B_0, R_0)$, either side can start ahead and lose, or start behind and win, as is observed historically.[19]

The most celebrated result of the theory is the so-called Lanchester Square Law, which is obtained easily. From (2.1), we have

$$\frac{dR}{dB} = \frac{bB}{rR}. \tag{2.3}$$

Separating variables and integrating from the terminal values $(R(t), B(t))$ to the higher initial values,

$$r\int_{R(t)}^{R_0} RdR = b\int_{B(t)}^{B_0} BdB,$$

we obtain the state equation

$$r(R_0^2 - R(t)^2) = b(B_0^2 - B(t)^2). \tag{2.4}$$

Of course, stalemate occurs when $B(t) = R(t) = 0$, which yields the Lanchester Square Law:

$$\begin{aligned} bB_0^2 &= rR_0^2 \quad \text{or} \\ B_0 &= \sqrt{\frac{r}{b}}R_0. \end{aligned} \tag{2.5}$$

This equation is very important. It says that, to stalemate an adversary three times as numerous, it does not suffice to be three times as effective; you must be nine times as effective! This presumed heavy advantage of *numbers* is deeply embedded in virtually all Pentagon models. For decades, it supported the official dire assessments of the conventional balance in Central Europe, giving enormous weight

[19] Indeed, the numerically smaller force was the victor in such notable cases as Austerlitz (1805); Antietam (1862); Fredericksburg (1862); Chancellorsville (1863); the Battle of Frontiers (1914); the fall of France (1940); the invasion of Russia (Operation Barbarossa, 1941); the battle of Kursk (1943); the North Korean invasion (1950); the Sinai (1967); the Golan Heights (1967 and 1973); and the Falklands (1982), to name a few.

to sheer Soviet numbers and placing a huge premium on western technological supremacy. That, of course, had budgetary implications. But, the presumption of overwhelming Soviet *conventional* superiority also shaped the development of so-called theater-nuclear weapons and produced a widespread assumption that their early employment would be inevitable, which drove the Soviets to seek preemptive offensive capabilities, and so on, in an expensive and dangerous military coevolution (see the preceding lecture).

The whole dynamic, while driven by myriad political and military-industrial interests on all sides, was certainly supported by Lanchester's innocent-looking *linear* differential equations (2.1). But, the linearity itself implicitly assumes things that are implausible on reflection and it mathematically precludes phenomena that, in fact, are observed empirically. Moreover, anyone exposed to mathematical biology would have found the Lanchester variant (2.1) to be suspect immediately.

## DENSITY

The equations, once again, are

$$\frac{dB}{dt} = -rR, \tag{2.6}$$

$$\frac{dR}{dt} = -bB. \tag{2.7}$$

In this framework, increasing density is a *pure benefit.* If the Red force $R$ grows, a greater volume of fire is focused on the Blue force $B$, and in (2.6), the Blue attrition rate grows proportionally. At the same time, however, *no penalty* is imposed on Red in (2.7) when, in fact, if the battlefield is crowded with Reds, the Blue target acquisition problem is eased and Red's attrition rate should grow.

In warfare, each side is at once *both predator and prey.* Increasing density is a benefit for an army as predator, but it is a cost for that same army as prey. The Lanchester square system captures the predation benefit but completely ignores the prey cost of density. The latter, moreover, is familiar to us all. For instance, if a hunter fires his gun into a sky black with ducks, he is bound to bring down a few. Yet if a single duck is flying overhead, it takes extraordinary accuracy to shoot it down. For ducks, considered as prey, density carries costs.

And, as any ecologist would expect, the effect is indeed observed. Quoting Herbert Weiss, "the phenomenon of losses increasing with force committed was observed by Richard H. Peterson at the Army Ballistic Research laboratories in about 1950, in a study of tank battles. It was again observed by Willard and the present author [Weiss] has noted its appearance in the Battle of Britain data."[20] The work referred to is D. Willard's statistical study of 1500 land battles.[21]

To his credit, Lanchester actually offered a second, nonlinear variant of these equations, which is much more plausible in this ecological light. Here,

$$\frac{dR}{dt} = (-bB)R\,, \tag{2.8}$$

$$\frac{dB}{dt} = (-rR)B\,. \tag{2.9}$$

In parentheses are the Lanchester square terms reflecting the "predation benefit" of density, but they are now multiplied by a term (the prey force level) reflecting "prey costs," as it were. The Red attrition rate in (2.8) slows as the Red population goes to zero, reflecting the fact that, as the prey density falls, the predator's search ("foraging") requirements for the next kill increase. Equivalently, Red's attrition rate grows if, like the ducks in the analogy, its density grows. In summary, a density cost is present to balance the density benefit reflected in the parenthesized term.

If we now form the casualty-exchange ratio

$$\frac{dR}{dB} = \frac{b}{r}\,,$$

separate variables, and integrate as before, we obtain the state equation

$$r(R_0 - R(t)) = b(B_0 - B(t))\,,$$

and the stalemate requirement

$$rR_0 = bB_0\,.$$

Now, as against the Lanchester Square Law, it *does* suffice to be three (rather than nine) times as good to stalemate an adversary three times as numerous.

## AMBUSH AND ASYMMETRY

Further, asymmetrical, variants of the basic Lanchester equations have been devised. For example, the so-called ambush variant imputes the "square law" fire concentration capacity to one side (the ambushers) but denies it to the other (the ambushees). Here,

$$\frac{dB}{dt} = -rR\,,$$

$$\frac{dR}{dt} = -bBR\,,$$

[20] Weiss (1966).

[21] Willard (1962).

so that

$$\frac{dB}{dR} = \frac{r}{bB},$$
$$b(B_0^2 - B(t)^2) = r(R_0 - R(t)).$$

Now assuming a fight to the finish ($R(t) = B(t) = 0$) and equal firing effectiveness ($r = b$), a Blue force of $B_0$ can stalemate a Red force numbering $B_0^2$—a hundred can hold off ten thousand. It's Thermopolae.

## REINFORCEMENT

Thus far the discussion has concentrated on the dynamics of *engaged* forces. Often, however, there is some flow of reinforcements to the combat zone proper. But, there are limits to the number of forces one can pack into a given area—there are "force to space" constraints. One might therefore think of the combat zone as having a carrying capacity and, accordingly, posit logistic reinforcement. Attaching such a term to the Lanchester nonlinear attrition model produces

$$\begin{aligned}\frac{dR}{dt} &= -bRB + \alpha R\left(1 - \frac{R}{K}\right),\\ \frac{dB}{dt} &= -rBR + \beta B\left(1 - \frac{B}{L}\right),\end{aligned} \tag{2.10}$$

where $\alpha, \beta, K$, and $L$ are positive constants. As observed in the preceding lecture, this is *exactly* Gause's (1935) famous model of competition between two species, itself a form of the general Lotka-Volterra ecosystem equations.

Equations (2.10) admit four basic cases, corresponding to different "war histories." These are shown in the phase portraits in figure 2.1.

FIGURE 2.1 Phase portraits for Lanchester/Gause Model

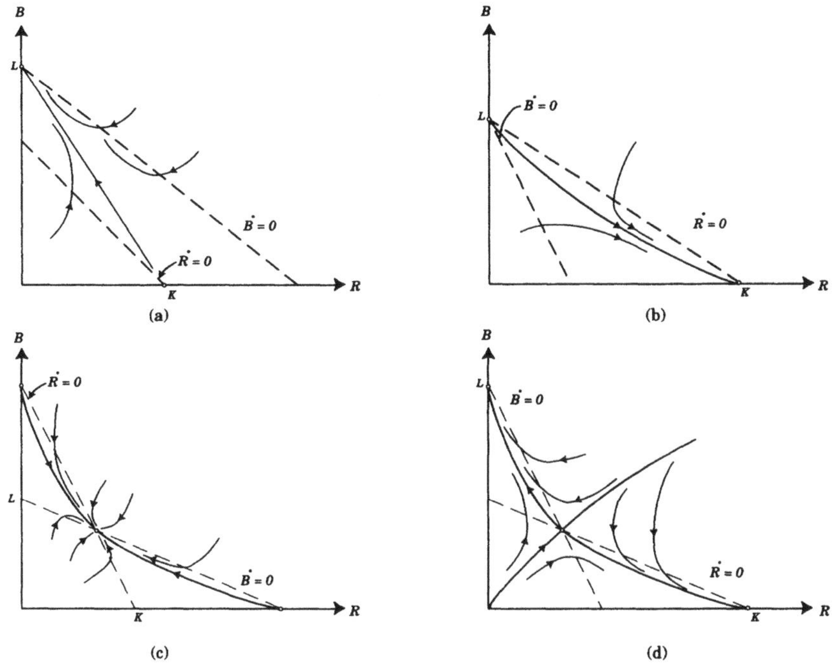


Source: Based on Clark (1990, p. 194).

Cases (a) and (b) are clear instances of the biological "principle of competitive exclusion," or military principle that one or the other side usually wins. Case (c) shows the horrific stable node—the "permanent war" that neither side wins. Finally, we have case (d), a saddle equilibrium. Any perturbation (off the stable manifold) sends the trajectory to a Red or Blue triumph. There is, however, the interesting and important region below both isoclines. Each side feels encouraged in this zone; reinforcement rates exceed attrition rates so the forces are growing. But, for instance, as the trajectory crosses the $\dot{B} = 0$ isocline, matters start to sour for Blue; $\dot{B}$ goes negative while Red forces continue to grow. Expectations of Blue defeat may set in, Blue morale may collapse, and, as a result, the Blue force can "break" long before it is physically annihilated. Indeed, the general phenomenon of "breakpoints" is common.

## BREAKPOINTS

Literal fights to the finish are actually rare. Normally, there is some level of attrition at which one belligerent "cracks." Suppose Blue breaks if $B(t) = \beta B_0$ and Red breaks if $R(t) = \rho R_0$, with $0 < \rho, \beta \leq 1$ and $\rho$ not necessarily equal to $\beta$. Clearly, breakpoints divide phase space into four zones, as shown in figure 2.2.

In Zone III, each side exceeds its breakpoint, so there is combat. Red wins if a trajectory crosses from Zone III to Zone II. All's quiet in Zone I, and so forth.

FIGURE 2.2 Breakpoints

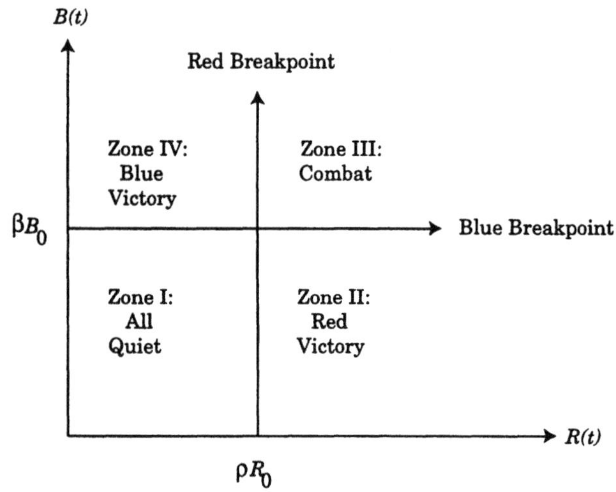


Substituting the stalemate conditions, $B(t) = \beta B_0$ and $R(t) = \rho R_0$ into, for illustration, the Lanchester square state equation (2.4) yields

$$r\left[R_0^2 - (\rho R_0)^2\right] = b\left[B_0^2 - (\beta B_0)^2\right],$$

which implies the (with breakpoints) stalemate condition

$$R_0\sqrt{r(1-\rho^2)} = B_0\sqrt{b(1-\beta^2)}.$$

## GENERALIZED EXCHANGE RATIO

As discussed in Epstein[22] these variants are all special cases of the general system

$$\frac{dR}{dt} = -bB^{c_1}R^{c_2}\,, \tag{2.11}$$

$$\frac{dB}{dt} = -rR^{c_3}B^{c_4}\,. \tag{2.12}$$

The corresponding casualty-exchange ratio is

$$\frac{dR}{dB} = \frac{b}{r}\frac{B^{c_1-c_4}}{R^{c_3-c_2}}\,,$$

where $c$-values are simply reals in the closed interval $[0, 1]$.

Clearly, from (2.11), $c_1$ is Blue's predation benefit from increasing density while, from (2.12), $c_4$ is Blue's prey cost of increasing density. Hence the exponent $c_1 - c_4$ might be thought of as the *net predation benefit of increasing density*, which is net fire concentration capacity in Lanchester's sense. The Red exponent $c_3 - c_2$ is analogously interpreted. Therefore, let us define

$$\begin{aligned} \lambda_b &= \text{ Blue's net predation benefit } = c_1 - c_4\,, \\ \lambda_r &= \text{ Red's net predation benefit } = c_3 - c_2\,. \end{aligned}$$

Then

$$\frac{dR}{dB} = \frac{b}{r}\left(\frac{B^{\lambda_b}}{R^{\lambda_r}}\right)\,. \tag{2.13}$$

Again separating variables and integrating from terminal to (higher) initial values, we have

$$b\int_{B(t)}^{B(0)} B^{\lambda_b}dB = r\int_{R(t)}^{R(0)} R^{\lambda_r}dR\,.$$

With stalemate defined as $B(t) = R(t) = 0$, we obtain the stalemate condition

$$\begin{aligned} \frac{b}{1+\lambda_b}B_0^{1+\lambda_b} &= \frac{r}{1+\lambda_r}R_0^{1+\lambda_r}\,, \text{ or} \\ B_0 &= \left[\frac{r}{b}\left(\frac{1+\lambda_b}{1+\lambda_r}\right)R_0^{1+\lambda_r}\right]^{\frac{1}{1+\lambda_b}}\,, \end{aligned} \tag{2.14}$$

which specializes to all the cases discussed earlier (e.g., $\lambda_b = \lambda_r = 1$ implies square law), and many more.

[22] Epstein (1985 and 1990).

Equation (2.13) is the algebraic form of the exchange ratio $\rho(t)$, used in my own Adaptive Dynamic Model. On separation of variables and integration, it also yields the measure of net military advantage used in the nonlinear arms race models of lecture 3.[23]

Of course, mere casualty-exchange ratios do not necessarily determine actual outcomes. Even defenders with favorable exchange ratios in engagements may run out of room or run out of time (e.g., popular support may collapse before the attacker's breakpoint is reached). Duration and territory—space and time—can loom every bit as large as physical attrition in determining outcomes. And this brings us to the topic of movement.

## MOVEMENT

Historically, war has been about territory. On a map of the modern world, the jagged borders are often simply the places where battle lines finally came to rest. It is interesting to compare these with the straight borders arrived at more contractually, peacefully—say, the borders between Nebraska and Kansas or between the U.S. and Canada. This is the reason that mountain ranges are such common borders: they were natural lines of military defense. The Alps, Himalayas, Pyrenees, and Caucases are examples. The same obviously holds for major bodies of water, like the English channel, and rivers, like the Yalu. In short, political borders reflect military technology. In any event, movement is a central aspect of war. And, as I argued at the outset, a plausible model should capture the basic connection between the fundamental processes: attrition and movement.

Lanchester himself had nothing to say about this and offered no model of movement. Contemporary extensions of Lanchester all handle it in essentially the same way: they posit that the velocity of the front—that is, the rate of defensive withdrawal—is some function of the force ratio. So, if

$$x \equiv \frac{R(t)}{B(t)},$$

these models posit a withdrawal rate, a velocity, $W(x)$ with $W(1) = 0$; $W'(x) > 0$ if $x > 1$ and eventually $W''(x) < 0$, implying some asymptote. (The direction of movement is always "forward" for the larger force.) One published example[24] is

$$W(x) = \frac{W_{\max}}{\sqrt{e^{(4/x)^2}}}.$$

[23] For further discussion of the $\lambda$'s, see Epstein (1990).

[24] Kaufmann (1983, p. 214).

The basic setup, then, is this: forces grind each other up via the attrition equations; the force ratio changes accordingly; and, as a function of that changing force ratio, the front's velocity changes, as depicted in the flow diagram of figure 2.3.

FIGURE 2.3 Flow Diagram for the Standard Model

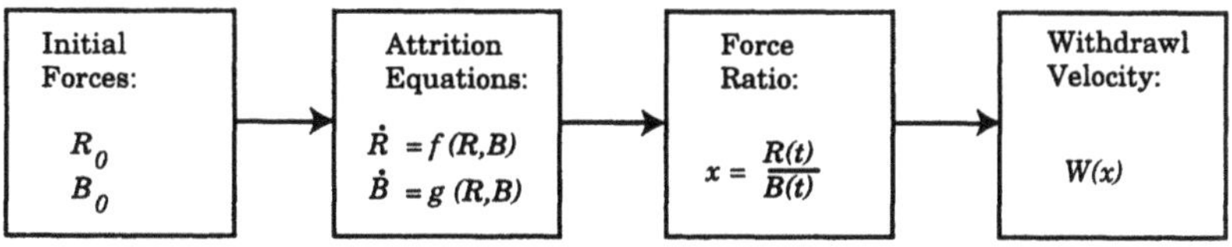


The framework is very neat indeed. The only problem is that any combat model with this basic structure is fundamentally implausible, and for one basic reason: movement of the front—defensive withdrawal—is anomalous! For a given pair of attacking and defending forces, the course of attrition on the defender's side, as calculated in this framework, is *exactly the same* whether he withdraws or not. The course of attrition on the attacker's side is also unchanged whether the defender withdraws or not. In short, defensive withdrawal neither benefits the defender nor penalizes the attacker. So, why in the world would the defender ever withdraw? The framework itself mathematically eliminates any rationale, or incentive, for the very behavior—withdrawal—it purports to represent. Movement is influenced by attrition, but not conversely. The movement of the front (withdrawal) is not *fed back* into the ongoing attrition process, when the entire point of withdrawal was presumably to affect that process—in the prototypical case, the point is to reduce one's attrition. Surely, it is contradictory to assume some benefit in withdrawal (otherwise, why would anyone withdraw?) and then to reflect *no benefit* whatsoever in the ongoing attrition calculations. Yet, all the contemporary Lanchester variants of which I am aware suffer this inconsistency.[25]

In turn, because defensive withdrawal cannot slow the defender's attrition (or, for that matter, the attacker's), the sacrifice of territory cannot prolong the war. And so, the most fundamental tactic in military history—the trading of space for time—is mathematically precluded. But, this tactic saved Russia from Napoleon and, later, from Hitler. A plausible model should certainly permit it.

[25]It is interesting to note that the battle of Iwo Jima—an island, where movement of the front was all but impossible—is the only case (to my knowledge) in which there is any statistical correspondence between events as they unfolded and as hypothesized by the Lanchester equations. Even if the statistical fit were good, there would be no basis for extrapolation to cases where substantial movement is possible. And, in fact, the fit is marred by insufficient data. On this issue, see Epstein (1985).

## THE ADAPTIVE DYNAMIC MODEL[26]

So, how do I fix it—how do I build in a *feedback* from movement *to* attrition? As simply as possible. The key parameters are the "equilibrium" attrition rates, $\alpha_{dT}$ and $\alpha_{aT}$. The first, $\alpha_{dT}$, is defined as the daily attrition rate the defender is willing to suffer in order to hold territory. The second, $\alpha_{aT}$, is defined as the daily attrition rate the attacker is willing to suffer in order to take territory. I assume $0 < \alpha_{dT}, \alpha_{aT} < 1$.

War, in addition to being a contest of technologies, is a contest of wills. So it is not outlandish to posit basic levels of pain (attrition rates) that each side comes willing to suffer to achieve its aims on the ground. If the defender's attrition rate is less than or equal to $\alpha_{dT}$, he remains in place. If his attrition rate exceeds this "pain threshold," he withdraws, in an effort to restore attrition rates to tolerable levels, an effort that may fail dismally depending on the adaptations of the attacker, a similar creature. If the attacker's attrition rate exceeds tolerable levels, he cuts the pace at which he prosecutes the war; if his attrition rate is below the level he is prepared to suffer, he increases his prosecution rate.[27]

It is the interplay of the *two adaptive systems, each searching for its equilibrium, that produces the observed dynamics, the actual movement that occurs and the actual attrition suffered by each side*. Indeed, in its most basic form, withdrawal might be thought of as an attrition-regulating servomechanism. The pain thresholds $\alpha_{dT}$ and $\alpha_{aT}$ play the roles of homeostatic targets, in other words. The introduction of these thresholds struck me—and still strikes me—as the most direct mathematical way to *permit* defensive withdrawal to affect attrition and, thus, to permit the trading of space for time. Their introduction also generates the fertile analogy between armies and a broad array of goal-oriented, feedback-control (cybernetic) systems.

Before delving into the mathematics, one possible misconception about these "pain" thresholds should be addressed. I do not claim, nor does my model imply, that battlefield commanders are necessarily *aware* of the numerical values of $\alpha_{dT}$ and $\alpha_{aT}$. Humans in the sixteenth century were not "aware" that they were sweating and shivering depending on the error: "body temperature minus 98.6 degrees, Fahrenheit." But the homeostatic behavior was there nonetheless.

## OVERVIEW OF THE MODEL

Let me now turn to the Adaptive Dynamic Model itself. The full apparatus includes air power as well as air and ground reinforcements, factors I will not discuss

[26] For earlier versions see Epstein (1985, 1990).

[27] These parameters represent daily *rates* of attrition, not total or cumulative attrition *levels*, as discussed above in connection with breakpoints.

here.[28] The model is a system of delay equations where the unit of time is usually interpreted as the day. If $A(t)$ and $D(t)$ are the attacker's and defender's ground forces surviving at the start of the $t$th day and $\alpha_a(t-1)$ is the attacker's attrition rate over the preceding day, we have the accounting identity

$$A(t) = A(t-1) - \alpha_a(t-1)A(t-1) \,. \tag{2.15}$$

The attacker's force on Tuesday is his force on Monday, minus total losses Monday. Likewise, it must be true that

$$D(t) = D(t-1) - (\text{Defender's losses on day}(t-1)) \,.$$

What are these losses? Well, if we define the casualty-exchange ratio as

$$\rho(t-1) \equiv \left( \frac{\text{Attackers Lost on day } t-1}{\text{Defenders Lost on day } t-1} \right) ,$$

the defender's losses must be

$$\frac{\alpha_a(t-1)A(t-1)}{\rho(t-1)} ,$$

since the numerator is the attackers lost on $(t-1)$. Thus, we have the second accounting identity

$$D(t) = D(t-1) - \frac{\alpha_a(t-1)A(t-1)}{\rho(t-1)} \,. \tag{2.16}$$

Obviously, once we attach specific functional forms to $\alpha_a(t)$ and $\rho(t)$, we no longer have accounting identities; we have a model. Above we discussed $\rho(t)$ and argued that a plausible and relatively general functional form is

$$\rho(t) = \rho_0 \frac{D(t)^{\lambda_d}}{A(t)^{\lambda_a}} , \tag{2.17}$$

where $\lambda_a, \lambda_d \in [0,1]$ are parameters. The real action—all feedback from movement to attrition—is inside $\alpha_a(t)$. Here is where the *interplay of adaptive belligerents* unfolds. As mentioned, this interplay is between the attacker's prosecution rate (reflecting the pace at which he chooses to press the attack) and the defender's withdrawal rate, both of which are attrition-regulating servomechanisms, in effect. The defender is, in some respects, simpler. We discuss him first.

[28] See Epstein (1990, pp. 85–99).

## ADAPTIVE WITHDRAWAL AND PROSECUTION

The defender's withdrawal rate for day $t$ is assumed to depend on the difference between his actual and his equilibrium attrition rate for the preceding day, day $(t-1)$. The functional form of that dependence should satisfy some basic requirements:

1. As the actual attrition rate for day $(t-1)$ approaches 1, the withdrawal rate for day $t$ should approach the maximum feasible daily rate, $W_{\max}$.
2. If the actual attrition rate for day $(t-1)$ is greater than the equilibrium rate $\alpha_{dT}$, the withdrawal rate for day $t$ should be greater than for day $(t-1)$.
3. If the actual attrition rate for day $(t-1)$ is less than or equal to the equilibrium rate $\alpha_{dT}$, then the withdrawal rate for day $t$ is zero.

It may not be correct, but the simplest functional form I can think of that satisfies these requirements is

$$W(t) = \begin{cases} 0 & \text{if } \alpha_d(t-1) \leq \alpha_{dT} \\ W(t-1) + \left(\frac{W_{\max} - W(t-1)}{1-\alpha_{dT}}\right)(\alpha_d(t-1) - \alpha_{dT}) & \text{otherwise} \end{cases}, \tag{2.18}$$

where

$$\alpha_d(t) = \frac{D(t) - D(t+1)}{D(t)}. \tag{2.19}$$

While in particular cases, there may be departures, exceptions,[29] and so forth, as a first-order idealization, the notion that, *ceteris paribus*, the aim of withdrawal is to reduce one's attrition rate seems fairly compelling. It also enjoys a certain biological plausibility. If the heat is too great, we yank our hand from the fire; Ashby's cat comes to mind. Surely, flight is a basic mechanism of defense for all species. One of the more famous experiments in this connection was conducted by our friend Gause and is known as his "flour beetle" experiment. He began with two beetle species competing in an environment of flour. He found competitive exclusion to be operative; left alone, one species consistently exterminated the other. But, when Gause inserted small lengths of glass tubing into the flour, the weaker species was able to retreat into the tubing, establish refuges, and survive—they could "trade space for time," as it were. So can the defenders in the Adaptive Dynamic Model. As we will see, they may choose to forego that option. But, a reasonable model should not preclude it.

Turning to the attacker, the model assumes that the pace at which he presses the attack, his prosecution rate for day $t$, which we denote $P(t)$, depends on the difference between his actual and his equilibrium attrition rates for the preceding

[29] Epstein (1985).

day, day $(t-1)$. The functional form of that dependence should satisfy some basic requirements:

1. As the attacker's actual attrition rate for day $(t-1)$ approaches 1, the prosecution rate for day $t$ should approach zero.
2. If the actual attrition rate for day $(t-1)$ is greater than (less than) the equilibrium rate $\alpha_{aT}$, the prosecution rate for day $t$ should be less than (greater than) for day $(t-1)$.
3. If the actual attrition rate for day $(t-1)$ equals the target, or equilibrium, rate, then there is no change in the prosecution rate.

It may not be correct, but the simplest functional form I can think of that satisfies these requirements is[30]

$$P(t) = P(t-1) - \left(\frac{P(t-1)}{1-\alpha_{aT}}\right)(\alpha_a(t-1) - \alpha_{aT}) . \tag{2.20}$$

As I said earlier, it is the *interplay* of these adaptive agents that shapes the dynamics; they are linked in the formula for $\alpha_a(t)$, the attacker's attrition rate for day $t$. This functional form should satisfy some basic requirements:

1. *Ceteris paribus*, the higher is the attacker's prosecution rate, the higher should be his attrition rate;
2. *Ceteris paribus*, the higher is the defender's withdrawal rate, the lower should be the attacker's attrition rate.
3. As the defender's withdrawal approaches full flight ($W(t) \to W_{\max}$), the attacker's attrition rate should approach zero.

It may not be correct, but the simplest functional form I can think of that satisfies these requirements is

$$\alpha_a(t) = P(t)\left(1 - \frac{W(t)}{W_{\max}}\right) . \tag{2.21}$$

Once the initial conditions and parameter values are specified, these equations produce the dynamics. And, as noted above, it is the coadaptation of these agents, each searching for its equilibrium, that determines the actual movement that occurs and the actual attrition that is suffered by each side.

In a nutshell, the attacker makes an opening "bid" on the pace of war, the rate at which his own forces are consumed (of course, he can set his rate at zero by not attacking). He may want to press the attack at an extremely high pace and may be willing to suffer extremely high attrition rates, if—for operational, strategic, or

[30] I am grateful to Mike Sobel for pointing out to me that the functional form for $P(t)$ that I originally published in Epstein (1985) actually fails requirement 2. Subsequent to our discussion, I noticed that it also fails 1.

political reasons—a quick decision is paramount.[31] Via the casualty-exchange ratio (defenders killed per attacker killed), this imposes an attrition rate on the defender. The latter may elect to hold his position and accept this attacker-dictated rate, or he may choose to reduce his attrition rate by withdrawing at a certain speed.

The mathematical mechanism whereby the defender's withdrawal reduces his attrition is not obvious. From (2.16), the attacker's attrition rate over day $t$, $\alpha_a(t)$, produces, via the inverse exchange ratio $1/\rho$, a defensive attrition rate over day $t$, $\alpha_d(t)$. If this exceeds the defender's movement threshold $\alpha_{dT}$, then on the next day the defender withdraws at a rate $W(t+1)$. This action reduces (that is, feeds back negatively on) the *attacker's* attrition rate $\alpha_a(t+1)$. In turn, this decrease in the attacker's attrition rate produces (again via $1/\rho$) a reduction in the defender's attrition rate $\alpha_d(t+1)$, whose size relative to $\alpha_{dT}$ determines the rate of any subsequent withdrawal. If $\alpha_d(t+1)$ is less than $\alpha_{dT}$, no subsequent withdrawal occurs. The front then remains in place unless and until the attacker—by attempting to force the combat at his chosen pace—imposes on the defender an attrition rate exceeding his withdrawal threshold, and so on. One might think of the defender as an adaptive system, with withdrawal rates as an attrition-regulating servomechanism.[32]

All the while, the attacker, too, is adapting; the prosecution rate $P(t)$ is his servomechanism. Just as there is some threshold $\alpha_{dT}$ beyond which the defender will withdraw, so the attacker possesses an "equilibrium" attrition rate $\alpha_{aT}$. If on day $(t-1)$ he records an attrition rate exceeding $\alpha_{aT}$, the attacker reduces the pace at which he prosecutes the combat. If he records an attrition rate lower than $\alpha_{aT}$, he accelerates by raising $P(t)$. The magnitude of these changes in $P(t)$ approach zero if the attacker's attrition rate approaches $\alpha_{aT}$, the equilibrium rate. Each side's adaptation may damp or amplify, penalize or reward, the adaptation of the other.

The adaptations are perhaps more sophisticated than meets the eye. Specifically, a primitive type of learning can occur. Suppose that on Monday, the defender's attrition rate exceeds his threshold $\alpha_{dT}$ by some amount $X$. In response, the defender withdraws at a rate $W(t)$ on Tuesday. Suppose, however, that—because his own attrition rate on Monday was below his threshold $\alpha_{aT}$—the attacker increases his prosecution rate on Tuesday and that, as a result, the defender's attrition rate on Tuesday again exceeds his threshold by the same amount $X$. Only a defender

[31] As an operational matter, a quick decision can circumvent logistical problems that could prove telling in a prolonged war. Strategically, the attacker may seek a decision before the defense has a chance to mobilize superior industry, superior reinforcements, or superior allies. An attacker with unreliable allies of his own may seek a quick win lest they begin to defect. An attacker may also choose to press the attack at a ferocious pace to secure a decision before the defender's nuclear options can be executed. A classic strategy of states facing enemies on multiple fronts has been to win quickly through offensive actions on one front and then switch forces to the second.

[32] Since initially publishing these equations, I have discovered the combat model of Rashevsky (1947). Although it differs from my model in numerous ways, Rashevsky's model does posit that, *ceteris paribus*, defensive withdrawal should reduce the attrition rates of both attacker and defender, and—using a different mathematization—it incorporates the idea of a defensive withdrawal threshold.

unable to learn would withdraw at $W(t)$ again, since that rate *already failed* to solve his problem. A more deeply adaptive defender would withdraw at a rate greater than $W(t)$; in the Adaptive Dynamic Model, he does. To me, this makes a certain amount of biological sense. If walking slowly away from a swarm of attacking bees does not reduce the sting rate, we try jogging. If jogging doesn't reduce the sting rate, we run, and so on, until we are running as fast as we can ($W_{\max}$). Of course, in the bee case we actually are free to pick something close to $W_{\max}$ as a first "trial retreat rate" because we are not concerned with territorial sacrifice. Analogous points apply to the attacker and his learning behavior in adjusting his prosecution rate, $P(t)$, as we will illustrate in the simulations below.

By setting the two fundamental thresholds $\alpha_{dT}$ and $\alpha_{aT}$ in various ways, the model will generate a reasonable spectrum of war types—bellotopes—from the war of entrenched defense, à la Verdun, to guerrilla war. I will discuss the four extreme settings and then present some simulations.

CASE 1: $\alpha_{aT} \approx 1$. The British at the Somme (1916) offer perhaps the great example of an attacker with no apparent pain threshold. Considering the extraordinary pain involved, we can ask with Jack Beatty, "What made them do it?"

> "'It' was to march, in an orderly way, rank by rank, column by column, to their death. That is what 20,000 British soldiers did on July 1, most of them falling between 7:30 and 8:30 A.M., the taste of tea and bacon still fresh on their lips. They got out of their trenches and marched to their death, or to some other form of mutilation.... Methodically, these [German] gunners raked the British formations. Methodically new formations set out, were shot down in no-man's-land, were replaced by other formations, and so on, turn and turn about, through the long day" (Beatty, 1986, pp. 112–114).

Long indeed. Here, perhaps, is a case of $\alpha_{aT} \approx 1$. Along similar lines, one thinks of the fateful Argonne Forest offensive of 1918 and, in particular, of Pershing's order to "push ahead *without regard to losses* and without regard to the exposed condition of the flanks." Surely, for Pershing, $\alpha_{aT}$ was close to 1. And, as Beatty notes, "It is no wonder that the cemetery at Romagne-Sous-Montfaucon, deep in the Argonne, is the largest American military cemetery in Europe, containing the remains of 14,246 soldiers."[33]

[33] Beatty (1986).

CASE 2: $\alpha_{dT} \approx 1$. The defensive analogue of the British at the Somme is undoubtedly the French at Verdun, also in 1916—not a good year, as Beatty recounts:

> "The French rotated seven tenths of their army though the meat grinder of Verdun. A colonel's order to his regiment gives the death-heavy flavor of the battle: 'You have a mission of sacrifice.... On the day they want to, they will massacre you to the last man, and it is your duty to fall.' The losses on both sides were appalling—perhaps a million and a quarter casualties in all. (The *ossuaire* at Verdun is full of the bones of the 150,000 unidentified and unburied corpses.) In short, Verdun was a demographic catastrophe for France. Yet, following Pétain's famous order, '*Ils ne passeront pas!*' the French Army held Verdun for the ten months of the battle—an epic of courage and endurance but not of victory. The standoff of Verdun, in the words of Alistair Horne, 'was the indecisive battle in an indecisive war; the unnecessary battle in an unnecessary war; the battle that had no victors in a war that had no victors'" (Beatty, p. 117).

Perhaps this is the terrible stable node I spoke of above—the sink of all sinks and, I would argue, a case of $\alpha_{dT} \approx 1$.

CASE 3: $\alpha_{dT} \approx 0$. Diametrically opposed to the French at Verdun are guerrilla defenders; their withdrawal threshold $\alpha_{dT}$ is close to zero. In guerrilla wars, like Vietnam, larger "superior" forces seeking direct engagements find themselves frustrated by defenders who withdraw—"vanish into the brush"—at the slightest attrition, the extreme case of trading space for time. Indeed, the entire strategy of the guerrilla—his only real hope—is precisely to *prolong* indecisive hostilities until domestic support for the war disintegrates, as it did for the United States in Vietnam.

CASE 4: $\alpha_{aT} \approx 0$. The fourth and final "pure" variant is the case where the attacker's equilibrium rate $\alpha_{aT}$ is close to zero. The natural example here is the so-called "fixing operation." The classic case is where an attacker is attempting a concentrated breakthrough in some sector of the battle front. He wants to prevent the defender from shifting forces from neighboring sectors to reinforce the breakthrough sector. Standard procedure for the attacker is to "pin," or "fix," these neighboring defensive forces by applying some pressure, but not enough to incur serious losses.

By specializing these two parameters, $\alpha_{dT}$ and $\alpha_{aT}$, the model will produce the "pure" forms, shown in table 2.1, as well as myriad mixed cases.

TABLE 2.1 Adaptive Dynamic Model

| Thresholds | |
|---|---|
| $\alpha_{dT}$ | Defender's Threshold |
| $\alpha_{aT}$ | Attacker's Threshold |
| **Qualitative Range** | |
| $\alpha_{dT} \rightarrow 1$ | Trench War (Verdun) |
| $\alpha_{dT} \rightarrow 0$ | Guerrilla War |
| $\alpha_{aT} \rightarrow 1$ | The Somme |
| $\alpha_{aT} \rightarrow 0$ | Fixing Operations |

## TWO SIMULATIONS

For illustrative purposes, I offer two simulations representing mixed cases. The numerical settings are given in table 2.2. In the first, I posit a ferocious attacker, with an equilibrium attrition rate of $\alpha_{aT} = 0.6$. The defender's withdrawal threshold attrition rate is set at $\alpha_{dT} = 0.3$, respectably stalwart. Though not shown in figure 2.4, the forces are initially equal (at half a million). What coadaptive story, then, is this picture telling?

The attacker's opening "bid" on the pace of war, his opening prosecution rate, is $P(1) = 0.1$. At this low level, the resulting attrition rate for the attacker is well below the 0.6 level he is, in fact, prepared to suffer. And so, as shown, he begins raising his prosecution rate. But, this must climb to around 0.4 (on day 3) before it produces a defensive attrition rate above the defender's threshold of $\alpha_{dT} = 0.3$, which induces withdrawal.[34] Both curves then rise to day 6. In this phase, the defender's withdrawals (partial disengagements) are thwarting the attacker's effort to attain his "ideal" attrition rate of $\alpha_{aT} = 0.6$, so the attacker prosecutes with increasing vigor, which efforts induce successive withdrawals at increasing rates.

[34]The computer has simply connected the dots in these pictures.

TABLE 2.2 Numerical Settings for Figures 2.4 and 2.5.

| Variable | Setting Figure 2.4 | Setting Figure 2.5 |
|---|---|---|
| $\alpha_{aT}$ | 0.6 | 0.1 |
| $\alpha_{dT}$ | 0.3 | 0.1 |
| $P(1)$ | 0.1 | 0.2 |
| $A(1)$ | $5 \times 10^5$ | —[1] |
| $D(1)$ | $5 \times 10^5$ | — |
| $W(1)$ | 0 | — |
| $W_{\max}$ | 20.0 | — |
| $\rho(t)$ | 1.1 | — |

[1] Dash indicates "same as in figure 2.4."

FIGURE 2.4 High Unequal Thresholds

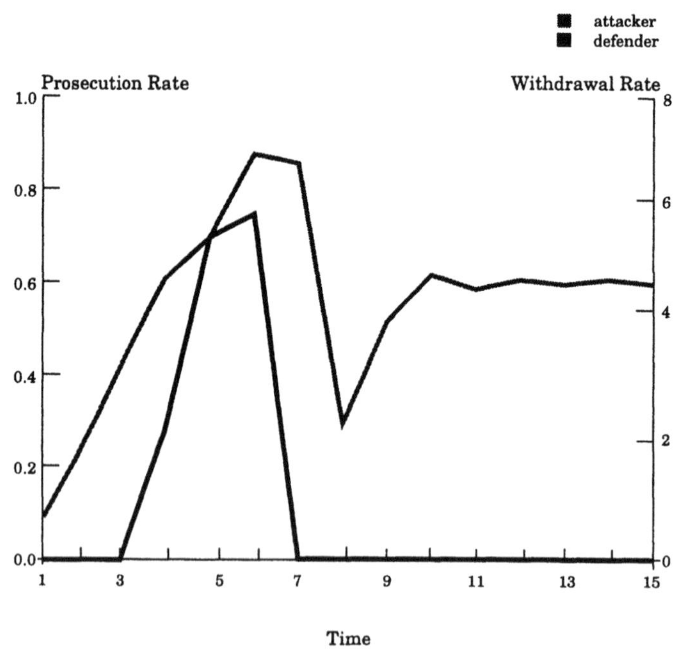

Now, all the while in this simulation, the casualty-exchange ratio (attackers killed per defender killed on day $t$) has been constant at a rate favoring the defender. And, by day 6, he has whittled down the attacker to such an extent that, even at high prosecution, the attacker cannot exact defensive attrition sufficient to induce withdrawal—so, withdrawal stops, the defender halts, on day 7.

In effect, the attacker "slams into" the now stationary defender on that day, producing attacker attrition well in excess of the attacker's tolerance $\alpha_{aT}$, "ouch," in other words. The attacker reacts to this extraordinary pain by cutting his prosecution rate sharply on day 8—too sharply, it turns out. He has overshot, as evidenced by his subsequent increases in $P(t)$ which ultimately levels off at around $P(t) = 0.6$.

A rather different history is portrayed in figure 2.5. The prosecution rate decreases monotonically, while the withdrawal rate rises and falls twice. In this case, the attacker's equilibrium, and defender's threshold, attrition rates are set equal at $\alpha_{aT} = \alpha_{dT} = 0.1$, considerably lower than in the preceding case. Initial force levels are as before.

FIGURE 2.5 Low Equal Thresholds

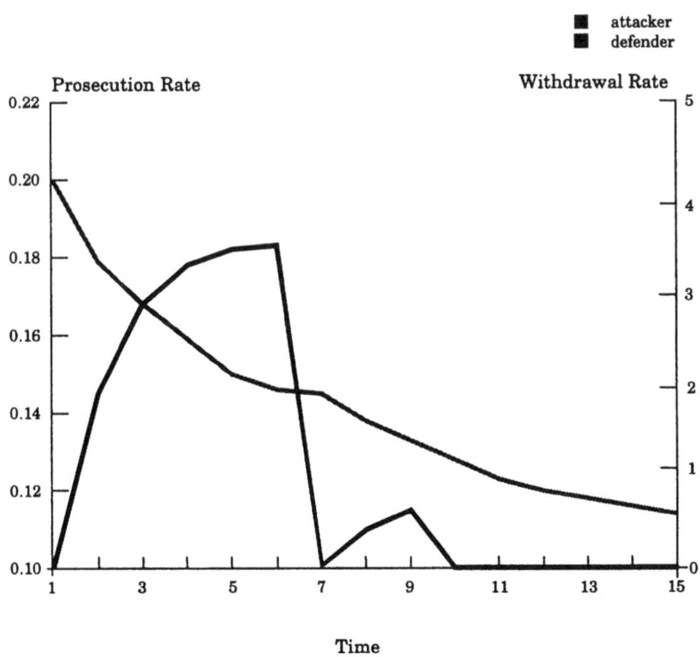

Here, the attacker's opening prosecution rate *exceeds* his equilibrium rate: $P(1) = 0.2$. This opening rate imposes on the defender an attrition rate that exceeds his withdrawal threshold. Over the first six days, *both* sides are above tolerance; the defender withdraws at a growing (though diminishing marginal) rate, while the attacker decreases his prosecution rate.

These coadaptations (plus a casualty-exchange ratio favoring the defender) gradually depress the defender's attrition rate to a level below his withdrawal threshold; so, on day 7, he halts. Though the attacker is steadily reducing his prosecution rate, the weight of his impact on the stationary defender is sufficiently painful to drive the latter from his position once more until, on day 10, the front stabilizes. The attack nonetheless persists, though at a declining level of ferocity, $P(t)$.

## SUMMARY

In Lanchester Theory—by which I mean the original equations and their contemporary extensions—these *behavioral* dimensions of combat are ignored. Mere opposing numbers and technical firing effectiveness completely determine the dynamics: there is no adaptation. In the Adaptive Dynamic Model, the parameters $\alpha_{dT}$ and $\alpha_{aT}$ allow one to reflect the different ways in which given forces can behave. As we have seen, with a given force, an attacker may prosecute the offensive at a ferocious pace, virtually unresponsive to losses. The British at the Somme in 1916 come to mind. Or, an attacker may operate *the same forces* at a more restrained pace, as in fixing operations. A high value of $\alpha_{aT}$ will produce the former type of attacker; a low value of $\alpha_{aT}$ will generate the latter.

Similarly, the tactical defender may be more or less stalwart in holding his positions. Guerrilla defenders may withdraw—"disappear"—when even slight attrition is suffered. For such tactical defenders, the withdrawal-threshold attrition rate $\alpha_{dT}$ is close to zero. At Verdun, by contrast, no attrition rate was high enough to dislodge the defenders from their entrenched positions. Pétain's famous order—"*Ils ne passeront pas!*"—effectively set $\alpha_{dT}$ equal to one.

These strategic and human realities are captured, however crudely, in the Adaptive Dynamic Model. And they are captured by a mechanism that permits movement to affect attrition, a *feedback* that is not possible in any version of Lanchester's equations. So, I feel some confidence in claiming that my equations present a *less crude* caricature of combat dynamics. But, given the complexity of the process, that is all I claim.

# LECTURE 3

# Imperfect Collective Security and Arms Race Dynamics: Why a Little Cooperation Can Make a Big Difference

This lecture uses simple mathematical models to explore the relationship between security regimes and arms race dynamics.[35] The main focus is a regime known as collective security, which is receiving wide attention. Little of the attention is mathematical, however, and, to my knowledge, none of it involves dynamical systems. One aim of this paper, then, is to formalize collective security in a dynamical systems context, which will allow us to extract some unexpected results. This formalization, of course, requires a rigorous definition of collective security. To wit: Imagine three countries $x$, $y$, and $z$. *Perfect* collective security would then operate as follows: If $x$ attacks $y$, $z$ allocates all force to $y$; if $y$ attacks $z$, $x$ allocates all force to $z$; and so on. The general rule is simply that *the odd man out instantly allocates all force to the attacked party.* In more biological—or sociobiological—terms, perfect collective security is a form of reciprocal altruism.[36]

[35] A regime is a "set of implicit or explicit principles, norms, rules, and decision-making procedures around which actors' expectations converge in a given area of international relations." See Krasner (1983, p. 2). On the emergence of norms generally, see Axelrod (1986).

[36] See Wilson (1975; 1978, chap. 7); Gould and Gould (1989, pp. 244–46); and Smith (1989, pp. 167–69).

Now, a heated debate surrounds this idea.[37] The debate turns on the question, "How much cooperation[38] is *possible*?" Skeptics argue that substantial levels of cooperation are not possible, and therefore that collective security can be dismissed. Proponents counter that a substantial degree of cooperation is possible and, therefore, that collective security is worth pursuing. Notice, however, that both positions *assume* substantial levels of cooperation to be *necessary* for collective security to be worthwhile. What about this central assumption? What about *a little bit* of collective security; is there merit in a highly diluted form? As nonlinear dynamicists—acutely aware that small perturbations can have huge effects—we are intrigued by the question.

And in fact, a central conclusion of this analysis is precisely that collective security regimes—even in highly diluted forms—can exert remarkably powerful stabilizing effects; in arms race models sufficiently nonlinear to produce really volatile dynamics, highly imperfect collective security regimes can damp the explosive oscillations and induce convergence to stable equilibria below initial armament levels. Put differently, the injection of *tiny* degrees of altruism can profoundly clam the otherwise volatile dynamics. The benefits of participating in the system are very great and, because of the nonlinearity, the required level of commitment from individual participants is very low.

In addition, one might assume that the more volatile a system is, the less value there will be in a given, low, level of collective security. But, counterintuitively within the class of models examined here, precisely the reverse is true! These results are examples of what I call "the nonlinear dynamics of hope," and would appear to invite a reorientation—or, at least, an extension—of research on altruism in a variety of fields.

## ALTRUISM: HOW LITTLE IS ENOUGH?

Specifically, the present analysis demonstrates that, in some dynamical systems, exceedingly low levels of *individual* altruism produce high levels of *collective* harmony. Everything depends on the intervening *dynamics*, and these, it seems to me, are largely missing from the debate. For instance, in a memorable phrase, Edward Wilson writes, "the genes hold *culture* on a leash"[39] (emphasis added). But, what he actually argues is that the genes hold *individuals* on a leash. The present analysis suggests that, even if the individuals' leash is very short, the culture's leash might be very long. Indeed, the analysis seems to open an emotionally appealing niche: perhaps we can be optimistic about the prospect of *social* harmony while

[37] A vigorous debate on the general topic has appeared in Harvard's journal, *International Security*. In particular, see Mearsheimer (1994/95), and Kupchan and Kupchan (1995). See also Kupchan and Kupchan (1991). For a collection of thoughtful analyses, see Downs (1994).

[38] Throughout, I will use the term "cooperation" and "reciprocal altruism," as just defined, interchangeably.

[39] Wilson (1978, p. 167).

retaining a certain degree of skepticism toward the prospect of *individual* altruism. In short, the issue is not simply how much individual altruism is *possible*; but, for social harmony, how little is *enough*?

## ORGANIZATION AND METHODOLOGY

The discussion is organized as follows. First, by way of introduction, the simplest two-country form of Lewis Frye Richardson's classic purely competitive arms race model is presented.[40] That model is then generalized slightly and expanded to encompass three competitors, the minimum number necessary to examine collective security.[41] In addition to Richardsonian competition and collective security, I will examine a regime characterized by the presence of a world policeman, or "globocop."[42] In this model, a force, $C$, located outside the competitive three-actor system is held at the ready to swoop in to support any attacked party. The object is to compare these three regimes formally. I do so first assuming that the underlying arms race dynamics are linear, as in Richardson's model.[43] Then I assume nonlinear arms race dynamics of a specific sort. I would hope that the full nonlinear model introduced here will contribute to the theoretical arms race literature in its own right.

Now, I make no attempt to test any model. Nor do I claim that any of these models is "right." Rather, the models are coarse lenses under which we compare the regimes. To be specific, the general procedure would be as follows. Take some model of Hobbesian—or, if you prefer, anarchic—arms race competition. Call that model $M_1$ (I use the linear Richardson model).[44] Then construct (see below) that model's collective security variant, $M_1^C$. And, for expository purposes, suppose collective security damps the competition in the sense that if $M_1^C(0) = M_1(0)$, then $M_1^C(t) < M_1(t)$ for all positive $t$.[45] Now, take a second Hobbesian model (I use a nonlinear Richardson model), $M_2$, construct its collective security variant, $M_2^C$, and compare the dynamics. Again, suppose that collective security damps the competition. Continue in this way. If this *comparative* result recurs without exception over a huge set of model pairs, $\{(M_i, M_i^C)\}$, then one may be justified in concluding that collective security exercises a systematically depressive effect on competitive

[40] Richardson (1960).

[41] I do not treat the $n$-actor, or globally inclusive, case here.

[42] In principle, globocop could be a consortium of powers. I thank Brian Pollins for this name.

[43] Assuming linearity, globocop is not an interesting variation on Richardson. In fact, it *is* linear Richardson with grievance terms translated by a fixed amount. Hence, little is said about globocop in Part I. Indeed, since its effects are, from a qualitative mathematical standpoint, pretty straightforward even in nonlinear cases, globocop is included for completeness but will receive relatively little attention.

[44] Technically, the model is affine.

[45] To spell this out completely, we are positing $M_1(t) \equiv (M_{11}(t), M_{12}(t), \ldots, M_{1n}(t))$ and $M_1^C(t) \equiv (M_{11}^C(t), M_{12}^C(t), \ldots, M_{1n}^C(t))$. Then $M_1^C(t) < M_1(t)$ iff $M_{1k}^C(t) < M_{1k}(t)$ for every $k$.

dynamics. This discussion suggests that may be the case, though more of this "structural sensitivity analysis" would be needed before confidence is obtained.[46]

From a methodological standpoint, it is also important to distinguish this analysis from other treatments of the issue. In particular, I am *not* examining the stability of collective security from a game theoretic standpoint.[47] No claim is made as to the likelihood of compliance with, or defection from, the system. Rather, I am trying to contribute a dynamical systems perspective to the theoretical literature, asking, with all else fixed, what is the dynamic effect of purely "institutional"—or, rule regime—change? The results bear on the game theoretic literature inasmuch as compliance depends on payoffs.[48] The payoffs associated with compliance—this analysis suggests—may be surprisingly high, and individual behavior may change as a result. Or, to couch it more prosaically, maybe if leaders appreciated the potential payoff of even limited collective security, they would be more interested in broad compliance. Indeed, the analysis would appear to raise starkly the question where to draw the line between altruism and self-interest in the international system.

Finally, I make no attempt to evaluate the practicality of implementing collective security in Europe, the Middle East, or any other particular region. However, if collective security in practice would behave at all like the idealized regime examined here, then its institution—even in highly diluted forms—might well be worth substantial effort. To begin at the beginning, let us revisit Richardson's original model.

[46] Ultimately, an elegant way to proceed would be to characterize mathematically the entire class of formal arms race models under which collective security (or globocop) would show a depressive effect, and then examine whether models falling outside that class are at all plausible. If not, we might wish to conclude that, by virtue of its membership in that class, the "right" model—whatever it is—will indicate the same depressive effect for collective security (or globocop), and *ipso facto*, that the effect is quite real. The basic idea, again, would be to say something reasonable about *comparative* dynamics, *without* claiming to know the "right" arms race model. Relatedly, it is worth stating explicitly that no sensitivity analysis on the parameter values given in the Appendix is conducted here. It is of considerable interest that there *exist* parameter settings at which a sharp sensitivity to rule regime is evident. A separate study would examine the robustness of this result under a wide range of parameter settings.

[47] See Niou and Ordeshook (1991). Also relevant are Axelrod (1984, 1987).

[48] The location of mixed strategy equilibria—and of evolutionarily stable strategies in bimatrix games—depends on payoff magnitudes, not just orderings. The speed of convergence to any stable equilibria also depends on payoff magnitudes.

# PART I. LINEAR MODELS

## THE CLASSIC RICHARDSON MODEL

The following differential equations constitute Richardson's basic model, with $x$ and $y$ the actors:

$$\dot{x} = a_2 y - a_1 x + g_x \,, \tag{3.1}$$

$$\dot{y} = b_1 x - b_2 y + g_y \,. \tag{3.2}$$

The basic idea is that a state's arms race behavior depends on three overriding factors: the perceived external threat, the economic burden of military competition, and the magnitude of grievances against the other party. Each merits a brief discussion.

## "GRIEVANCES"

The constants, $g_x$ and $g_y$, are usually interpreted simply as grievances. And, on this reading, a core message of Richardson's model is that there can be no permanent disarmament without the resolution of underlying political grievances. Even if disarmament is total (i.e., $x(t) = y(t) = 0$), arms racing will reemerge if $g_x$ or $g_y$ is greater than zero. For instance, take (3.1), and assume $x(t)$ and $y(t)$ are both zero. If $g_x > 0$, then, since the growth *rate* $\dot{x}(t)$ equals $g_x$ , that rate, too, must exceed zero. Hence $x(t)$ begins to grow, which, via (3.2), stimulates a growth reaction in $y(t)$, which feeds back to further stimulate $x(t)$, and the race is on. Or, as Richardson himself put it, "mutual disarmament without satisfaction is not permanent."[49]

It seems to me that $g_x$ and $g_y$ can be interpreted somewhat more broadly. States compete militarily not only because there are grievances, but also because they lack confidence that grievances can be resolved without resort to arms. The emergence of institutions offering high confidence that differences could be resolved nonviolently might permit disarmament despite outstanding grievances. So, I think of the terms, $g_x$ and $g_y$, as capturing both the underlying grievances and the level of confidence that grievances can be resolved nonmilitarily. If confidence ranges from zero to one, then each $g$ could be interpreted as: (grievance)·(1-confidence), for example. It is far from clear how one would measure $g_x$ or $g_y$ under either interpretation. Luckily, their measurement is not necessary for our purposes.

[49] Richardson (1960, p. 17).

## THE "ECONOMIC FATIGUE" TERM

Leaving the grievance terms aside, the rate at which $x$ grows, $\dot{x}$, is proportional to the perceived external threat $y$, and vice versa for the growth rate $\dot{y}$. Without some damping term, this is a pure positive feedback system that simply blows up. Richardson posited economic fatigue terms ($-a_1x$ and $-b_2y$) that damp the process. Clearly, so long as the fatigue coefficients $a_1$ and $b_2$ are greater than zero, the process is damped. Importantly, if $a_1$ and $b_2$ are negative, then $-a_1$ and $-b_2$ are positive, meaning that military growth rates increase the larger is the military establishment. In such cases, $x$ and $y$ may both grow even in the absence of underlying grievances or any perceived threat. This would be autocatalytic growth in the "military-industrial complex." The "fatigue" coefficients might be thought of as embodying the net civil-military, or "guns versus butter" balance in society. If the terms are negative, then there is auto-catalytic military expansion, or "militarism" for short.

## THE EXTERNAL THREAT TERMS

Finally, the terms $b_1x$ and $a_2y$ incorporate perceived external threats, obvious components of any plausible model of arms race behavior.[50] I will generalize this model below, adding a third party and introducing certain nonlinearities, among other things. But clearly, Richardson's linear model has some basic appeal, parsimoniously relating arms race behavior to grievances, perceived external threats, and internal economic fatigue.[51]

## SIMPLE ANALYTICS

The simple analytics of the famous model deserve a concise review. Defining

$$A = \begin{pmatrix} -a_1 & a_2 \\ b_1 & -b_2 \end{pmatrix},$$

the basic Richardson model is simply

$$\dot{x} = Ax + g; \quad x, g \in \mathcal{R}^2. \tag{3.3}$$

[50]In reality, these terms are probably nonlinear, since the military's power to *shape treat perceptions* itself may vary with the size of the military establishment, suggesting terms of the form $a_2(x)y$ and $b_1(y)x$. See Morel (1991).

[51]The model provides a nice framework for interpreting gross changes in, for instance, Soviet behavior. Thinking of the Soviets as country $x$, one might argue that the end of the cold war manifests the facts that Gorbachev's $a_2$ < Brezhnev's $a_2$, Gorbachev's $g$ < Brezhnev's $g$, and Gorbachev's $a_1$ > Brezhnev's $a_1$. The last of these—Gorbachev's assessment of the economic burden of the arms race—was perhaps crucial. But, in any event, he adjusted the Soviet parameters sharply and, suddenly, arms race dynamics were changed.

A positive equilibrium $\bar{x}$, if it exists, is given by

$$\bar{x} = -A^{-1}g\,.$$

Stability is independent of $g$; that is, $\bar{x}$ is a stable equilibrium of (3.3) if and only if the origin is a stable equilibrium of $\dot{z} = Az$, where $z = x - \bar{x}$. As covered in lecture 6, that requirement is met if Tr$A < 0$ and Det $A > 0$. For $a_1$ and $b_2$ positive, the trace condition is obviously satisfied by $A$, and the determinant is positive if the product of economic fatigue terms ($a_1b_2$) exceeds the product of reciprocal activation terms ($b_1a_2$). As observed in lecture 1, the global equilibrium of the Richardson model is the interior equilibrium of the mutualistic Lotka-Volterra ecosystem model.

Now, simple linear model in hand, we wish to examine the transition from this Hobbesian world to a collective security regime first and, secondarily, to a world characterized by the presence of a world policeman, or "globocop."

As noted at the outset, there has been no attempt to frame the comparison in dynamical systems terms. An immediate issue, then, is how to operationalize collective security and globocop. Recall that under a *perfect* (as against diluted) three-party collective security regime, if $x$ attacks $y$, $z$ instantly contributes all forces to $y$. If $z$ attacks $x$, $y$ instantly allocates all forces to $x$, and so forth. The odd man out instantly allocates all force to the attacked party. Under globocop, a force of fixed size, $C$, from outside the three-actor system is instantly allocated to any attacked party. Various imperfections will be discussed below. But *perfect* collective security, and globocop, are operationalized in figure 3.1, as variations on an underlying linear Richardson model.

There are clearly some differences between these systems and (3.1) and (3.2). First, rather than two actors, there are three, the minimum number necessary to examine collective security. Second, these are systems of difference, rather than differential, equations. For notational simplicity, $\Delta x \equiv x_{t+1} - x_t$, and unsubscripted state variables on the right-hand sides represent values at time $t$. From this point on, we will work in discrete time, mainly because major decisions on national armament levels are taken in discrete time (e.g., annually).[52]

Turning now to more substantive issues, notice that the Richardson model is reformulated slightly in that actors respond not simply to adversary military *levels*, as in (3.1) and (3.2), but to the *gaps* between their own levels and those of potential adversaries; country $x$ reacts to $(y - x)$ in the first equation rather than to $y$ as before. Here, with Richardson, we "suppose that what moves a government to arm is not the absolute magnitude of other nations' armaments but the difference between its own and theirs."[53]

[52] In fact, the discrete nonlinear dynamics (below) are considerably richer than the analogous continuous dynamics.

[53] Richardson (1960, p. 35).

FIGURE 3.1 Linear Models: The Three Basic Variants[54]

LINEAR RICHARDSONIAN COMPETITION.

$$\begin{aligned}\Delta x &= -a_1 x + a_2(y - x) + a_3(z - x) + g_x\\ \Delta y &= b_1(x - y) - b_2 y + b_3(z - y) + g_y\\ \Delta z &= c_1(x - z) + c_2(y - z) - c_3 z + g_z\end{aligned}$$

LINEAR AND PERFECT COLLECTIVE SECURITY.

$$\begin{aligned}\Delta x &= -a_1 x + a_2(y - (x + z)) + a_3(z - (x + y)) + g_x\\ \Delta y &= b_1(x - (y + z)) - b_2 y + b_3(z - (y + x)) + g_y\\ \Delta z &= c_1(x - (z + y)) + c_2(y - (z + x)) - c_3 z + g_z\end{aligned}$$

LINEAR GLOBOCOP.

$$\begin{aligned}\Delta x &= -a_1 x + a_2(y - (x + C)) + a_3(z - (x + C)) + g_x\\ \Delta y &= b_1(x - (y + C)) - b_2 y + b_3(z - (y + C)) + g_y\\ \Delta z &= c_1(x - (z + C)) + c_2(y - (z + C)) - c_3 z + g_z\end{aligned}$$

Under Richardson, when $x$ evaluates the external threat from $y$, he computes $y - x$. But under perfect collective security, he can count on $z$'s undiluted support. So, he computes $y - (x + z)$. In turn, when $x$ evaluates the potential threat posed by $z$, he computes $z - x$ under Richardson, but $z - (x + y)$ under perfect collective security. Likewise for all external threat terms in the model. Even this very simple formulation suggests that life under collective security would differ from life under Richardson in nonobvious ways. For instance, put yourself in $x$'s position and ask: am I better off or worse off if $y$'s arsenal grows? In a Richardsonian world, the answer is clear: you are worse off. Under collective security, by contrast, the answer is not clear since $y$ is a threat in one context but is an ally in another (namely, if $z$ attacks). Whether $x$ ultimately rises or falls with an increase in $y$ depends on the quantity $a_2 - a_3$, which might be positive, negative, or zero.

Relative to linear Richardson, however, linear perfect collective security has a systematically depressive effect on the competition. While this result is intuitive, a formal proof (below) will, in fact, lead to counterintuitive results. In particular, it is the *magnitude* of "the collective security effect"—not its sign—which is unexpected, particularly in the nonlinear variants below. But, having found a simple way to import collective security—or reciprocal altruism—into a basic arms race model, let us delay proofs and structural variations for some elementary simulations. Though

[54]Although equivalent matrix formulations will be used below, I eschew matrices here for two reasons. First, for the uninitiated, the basic differences between the regimes comes through more clearly with this notation. Second, and more important, the relationships between the linear versions and the corresponding nonlinear variations below will be much clearer in the notation.

informal, these can help us develop a "feel" for how collective security may effect dynamics in the perfect linear case.

For illustrative purposes, then, a Base Case simulation of the linear Richardson model is given in figure 3.2. All actors are set at 1000 units initially, and all reaction coefficients, economic fatigue coefficients, and grievances are assumed positive. All assumptions are given in the Appendix.

On those assumptions, the evolution is shown in figure 3.2. The defense levels all increase, leveling off to some equilibrium, $\bar{x} \in \mathcal{R}^3_+$, given by $\bar{x} = (I - M)^{-1}g$, for appropriately defined $M$ and $g$.[55]

FIGURE 3.2 Linear Richardson

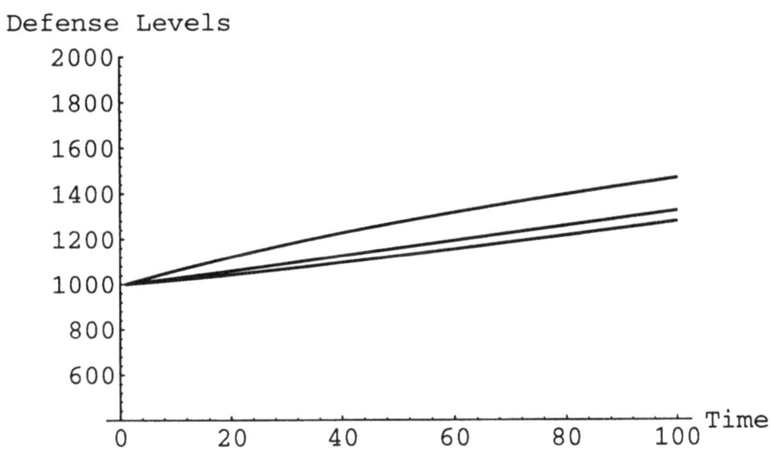


Now, leaving all initial defense levels and other numbers fixed, what is the effect of the purely *institutional* transition to collective security? How does the change in *rule regime* alter dynamics? Instead of growth, we have decline, as shown in figure 3.3.

[55] In matrix notation, all three linear models in figure 3.1 share the general form, $x_{t+1} = Mx_t + g$. By definition, an equilibrium, $\bar{x}$, satisfies $\bar{x} = M\bar{x} + g$ so, where it exists (i.e., where $I - M$ is nonsingluar), $\bar{x} = (I - M)^{-1}g$.

FIGURE 3.3 Linear Collective Security

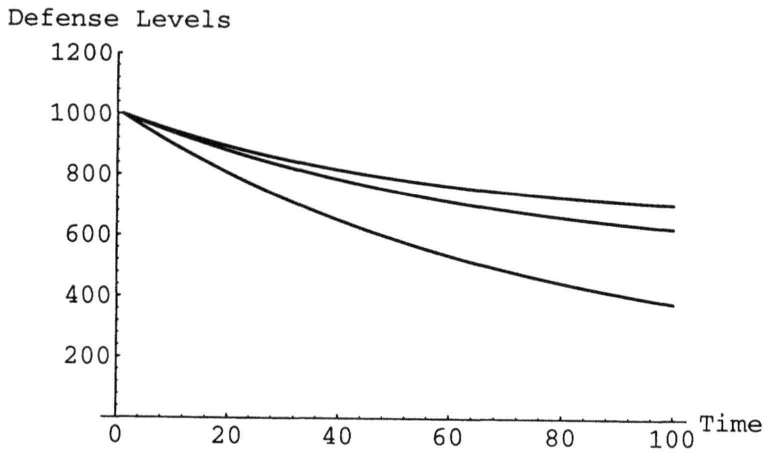


In fact, if all coefficients are positive, both systems will attain equilibrium (more on this below).

FIGURE 3.4 Richardson Autocatalytic

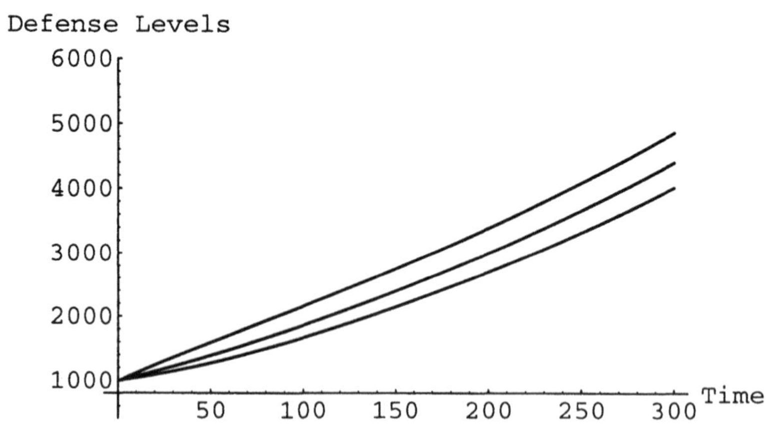

So powerful is the collective security effect, however, that even cases of autocatalytic arms growth ($a_1, b_2, c_3$ all negative), or "militarism," can be reversed. Figure 3.4 shows an autocatalytic (*negative* economic fatigue) case under Richardson.

FIGURE 3.5 Collective Security Autocatalytic

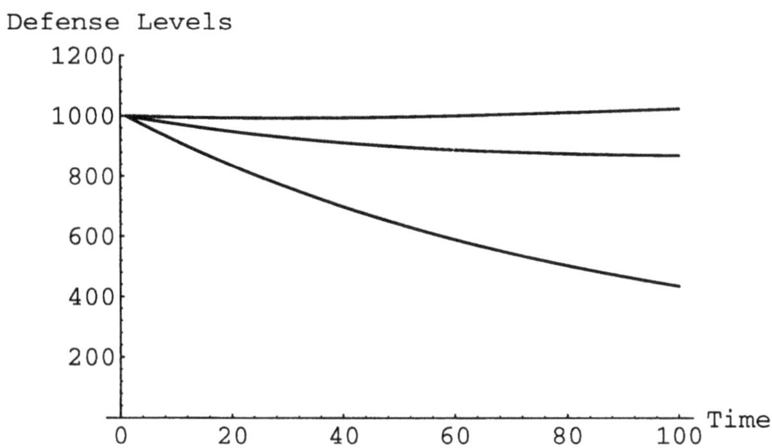


And, in figure 3.5, we have exactly the same numerical settings, but run under perfect collective security.

These results are systematic. Indeed, if we denote the Richardsonian, collective security, and globocop levels at time $t$ by $x_t^R, x_t^{CS}$, and $x_t^{GC}$, then

**Theorem.** *If* $x_0^R = x_0^{CS} = x_0^{GC} > 0$, *then for all positive* $t$, $x_t^R > x_t^{CS}$ *and* $x_t^R > x_t^{GC}$. *Here, for* $x, y \in \mathcal{R}^n$, $x > y$ *iff* $x_i > y_i$ *for all* $i$.

**Proof.** Casting the above systems in matrix form, let

$$A = \begin{pmatrix} 1-(a_1+a_2+a_3) & a_2 & a_3 \\ b_1 & 1-(b_1+b_2+b_3) & b_3 \\ c_1 & c_2 & 1-(c_1+c_2+c_3) \end{pmatrix},$$

$$b = \begin{pmatrix} g_x \\ g_y \\ g_z \end{pmatrix},$$

$$\nu = \begin{pmatrix} a_2+a_3 \\ b_1+b_3 \\ c_1+c_2 \end{pmatrix}, \text{ and}$$

$$B = \begin{pmatrix} 0 & a_3 & a_2 \\ b_3 & 0 & b_1 \\ c_2 & c_1 & 0 \end{pmatrix}.$$

For future reference, it is important to note that if $x > 0$, then $Bx > 0$.[56] With these definitions, it is a matter of trivial algebra to show that

$$x_{t+1}^{R} = Ax_t^R + b, \tag{3.4}$$
$$x_{t+1}^{CS} = (A-B)x_t^{CS} + b, \tag{3.5}$$
$$x_{t+1}^{GC} = Ax_t^{GC} + (b - C\nu). \tag{3.6}$$

If $x_0^R = x_0^{GC}$, then globocop (3.6) is simply Richardson (3.4) with grievances reduced by the constant vector $C\nu > 0$; the effect is obviously depressive. To prove that collective security is strictly depressive, we establish a simple lemma.

**Lemma.** If, for some time $\hat{t}, x_{\hat{t}}^R > x_{\hat{t}}^{CS}$, then $x_t^R > x_t^{CS}$ for all $t > \hat{t}$.

**Proof.**

$$\begin{aligned} x_{\hat{t}+1}^R &= Ax_{\hat{t}}^R + b && \text{by (3.4)} \\ &> (A-B)x_{\hat{t}}^R + b && \text{since } Bx_{\hat{t}}^R > 0 \\ &> (A-B)x_{\hat{t}}^{CS} + b && \text{since } x_{\hat{t}}^R > x_{\hat{t}}^{CS} \text{ by hypothesis} \\ &= x_{\hat{t}+1}^{CS} && \text{by (3.5)}. \end{aligned}$$

By this Lemma, a proof that $x_t^R > x_t^{CS}$ for all positive $t$ will be in hand once we show that $x_1^R > x_1^{CS}$. But this is simple. Since $x_0^R = x_0^{CS}$, call them both $x_0 > 0$. Then,

$$x_1^R - x_1^{CS} = [Ax_0 + b] - [(A-B)x_0 + b] = Bx_0 > 0,$$

[56]To ensure physical realism ($x > 0$), we must stipulate that $a_1 + a_2 + a_3 < 2, b_1 + b_2 + b_3 < 2$, and $c_1 + c_2 + c_3 < 2$. I thank Jean-Pierre Langlois for this observation.

and we are through. □

Now, as noted earlier, these results obtain so long as $Bx > 0$ for $x > 0$. The *specific* $B$ matrix above can be altered considerably while leaving the strictly depressive effect of collective security intact. Connectionist terminology will prove natural for discussing imperfect collective security.

## THE CONNECTIONISM OF COLLECTIVE SECURITY

This perspective emerges from closer scrutiny of the $B$ matrix. The $ij$th entry, $b_{ij}$, represents the level of altruism that party $j$ shows party $i$. If we let the symbol "$x \rightarrow y$" represent the altruism $x$ shows $y$ (i.e., it is $b_{21} > 0$), then, conceptually, the $B$ matrix is

$$B = \begin{pmatrix} 0 & y \rightarrow x & z \rightarrow x \\ x \rightarrow y & 0 & z \rightarrow y \\ x \rightarrow z & y \rightarrow z & 0 \end{pmatrix}.$$

Graphically, this would correspond to the "altruism web" shown in figure 3.6. Pairwise, all altruism is reciprocated; arrows run in both directions. If $z$ attacks $x$, $y$ allocates force to $x$ and vice versa if $z$ attacks $y$, and so on. When this is the case, we will say that the collective security system is *maximally connected.*

FIGURE 3.6 Maximally Connected Altruism Web

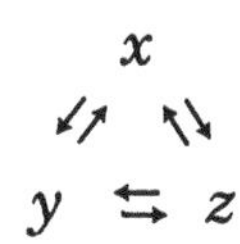


All off-diagonal elements of the $B$-matrix are strictly positive; the sign structure is then

$$B = \begin{pmatrix} 0 & + & + \\ + & 0 & + \\ + & + & 0 \end{pmatrix}.$$

The *strength* of any connection (in contrast to the connection *pattern*) is the real number, $b_{ij}$, which can assume values in $[0, 1]$. So, in these terms, perfect collective security entails maximum connectivity and maximum connection strength. In turn, imperfect collective security regimes result from reductions in connectivity, reductions in connection strength, or both.

## MAXIMAL CONNECTIVITY WITH DILUTED STRENGTH

It is obvious that, if $Bx > 0$ then $\gamma Bx > 0$ for any real $\gamma \in (0,1)$. This is the most transparent case of imperfect or "diluted" collective security. Maximal connectivity—reciprocal altruism—prevails, but, instead of sending *all* of one's forces to the aid of the attacked party, one sends a fraction $\gamma$. In fact, the strictly depressive effect is preserved if every off-diagonal entry in the $B$-matrix is a different $\gamma_{ij} \in (0,1)$. Everyone is better off even if the reciprocal altruism is, in this sense, discriminatory.

## MINIMAL CONNECTIVITY WITH DILUTED STRENGTH

More intriguing, however, the altruism need *not* be reciprocal to leave all parties strictly better off. Specifically, altruism matrices far more sparse than $B$ will fulfill our strictly depressive requirement, $Bx > 0$. Indeed, it is necessary only that each row contain a *single* positive entry. So, for instance, any matrix with the following sign structure will do.

$$B = \begin{pmatrix} 0 & 0 & + \\ + & 0 & 0 \\ 0 & + & 0 \end{pmatrix}.$$

Graphically, this would correspond to the cyclic "altruism web" in figure 3.7.

FIGURE 3.7 Cyclic Altruism Web

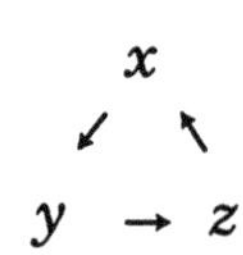


Here, $x$ is unilaterally altruistic to $y$; $y$ is unilaterally altruistic to $z$; and $z$ is unilaterally altruistic to $x$. Everyone is better off, but there is *no reciprocal altruism.* Instead of "you scratch my back and I'll scratch yours," the appeal is "you scratch my back, and I'll scratch Sam's, and Sam will scratch yours." I call this "cyclic altruism."[57] The direction of the cycle is reversed if $B$ has the sign pattern shown below.

$$B = \begin{pmatrix} 0 & + & 0 \\ 0 & 0 & + \\ + & 0 & 0 \end{pmatrix}.$$

[57] Obviously, this is a form of diluted altruism, with some $\gamma_{ij}$'s equal to zero. But, as it has a different flavor and, because the *position* of the zeros matters, I give it a separate name.

It is, in fact, not necessary that these altruism graphs, or "webs," be closed. For instance, any $B$ matrix with the following sign pattern will satisfy our strictly depressive, $Bx > 0$, requirement.

$$B = \begin{pmatrix} 0 & + & 0 \\ + & 0 & 0 \\ 0 & + & 0 \end{pmatrix}.$$

But, its graph is open, as shown in figure 3.8.

FIGURE 3.8 Open Altruism Web

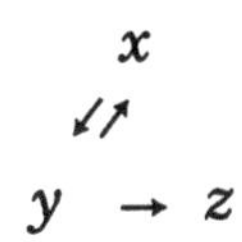


*Everyone* is strictly better off if $x$ and $y$ are reciprocal altruists and $y$ is unilaterally altruistic to $z$, even if $z$ is altruistic to no one!

In summary, for the linear models above, there are basically two senses in which collective security can be imperfect and still exert a strictly depressive effect on dynamics. Altruism can be perfectly reciprocal but diluted in strength. It can also be imperfectly reciprocal (as in figure 3.8), even unreciprocated (as in figure 3.7). As we will see, it may in fact be *both* highly diluted and imperfectly reciprocal and still have a profoundly depressive effect. It is in precisely the systems that concern us most—the volatile systems—that such highly imperfect collective security regimes can have dramatic stabilizing effects. Such dynamics, however, really arise only in *nonlinear* systems. Let us turn to these variants.

## PART II. NONLINEAR MODELS

Nonlinearities may enter the model in numerous ways. One way is through the balance assessment, or external threat, terms. Modern military establishments do not measure military balances—external threats—by simple subtractions of the form $y - x$, as in the above models. Rather, they often use methods that, at some level or other, embed mutual attrition models implying that *net military advantage is a difference of levels raised to powers.* Where does this come from? Basically, from the attrition stalemate conditions of generalized Lanchester equations which were discussed in the preceding lecture. Allow me to derive this quickly.

## GENERALIZED ATTRITION STALEMATE

Let $R(t)$ and $B(t)$ be Red and Blue forces at time $t$, and let $r$ and $b$ (real numbers between zero and one) represent their effectiveness per unit. With constants $c_1$ through $c_4 (0 \leq c \leq 1)$, the most general Lanchester attrition system is

$$\frac{dR}{dt} = -bB^{c_1}R^{c_2}\,,$$
$$\frac{dB}{dt} = -rR^{c_3}B^{c_4}\,.$$

These relations imply that the casualty-exchange ratio (Reds killed per Blue killed) is[58]

$$\frac{dR}{dB} = \frac{b}{r}\frac{B^{c_1-c_4}}{R^{c_3-c_2}}\,.$$

Assuming, for simplicity's sake, that $c_1 - c_4 = c_3 - c_2 = \lambda$, let us separate variables and integrate from terminal to (higher) initial values.

$$r\int_{R(t)}^{R(0)} R^{\lambda}dR = b\int_{B(t)}^{B(0)} B^{\lambda}dB\,.$$

With stalemate defined as $R(t) = B(t) = 0$, we obtain the stalemate condition:

$$\frac{1}{\lambda+1}[bB(0)^{\lambda+1} - rR(0)^{\lambda+1}] = 0\,.$$

If the left-hand side is greater than zero, Blue "wins." If it is less than zero, Red "wins." Assuming equal firing effectiveness $b = r$, and defining $\beta = \lambda + 1$, the measure of net military advantage implicit in all Lanchester-based attrition models is therefore

$$\frac{1}{\beta}(B^{\beta} - R^{\beta})\,,$$

which is exactly the form employed in the nonlinear Richardson, collective security, and globocop variants displayed in figure 3.9.

[58] For a fuller discussion of this generalized exchange ratio and its interpretation, see Epstein (1993, and 1990, pp. 85–93), and lecture 2 above.

FIGURE 3.9 Nonlinear Variants

NONLINEAR RICHARDSONIAN COMPETITION ($1 < \beta \leq 2$).

$$\Delta x = -a_1 x + \frac{a_2}{\beta}(y^\beta - x^\beta) + \frac{a_3}{\beta}(z^\beta - x^\beta) + g_x$$
$$\Delta y = \frac{b_1}{\beta}(x^\beta - y^\beta) - b_2 y + \frac{b_3}{\beta}(z^\beta - y^\beta) + g_y$$
$$\Delta z = \frac{c_1}{\beta}(x^\beta - z^\beta) + \frac{c_2}{\beta}(y^\beta - z^\beta) - c_3 z + g_z$$

NONLINEAR AND VARIABLE COLLECTIVE SECURITY.

$$\Delta x = -a_1 x + \frac{a_2}{\beta}(y^\beta - (x + \gamma z)^\beta) + \frac{a_3}{\beta}(z^\beta - (x + \gamma y)^\beta) + g_x$$
$$\Delta y = \frac{b_1}{\beta}(x^\beta - (y + \gamma z)^\beta) - b_2 y + \frac{b_3}{\beta}(z^\beta - (y + \gamma x)^\beta) + g_y$$
$$\Delta z = \frac{c_1}{\beta}(x^\beta - (z + \gamma y)^\beta) + \frac{c_2}{\beta}(y^\beta - (z + \gamma x)^\beta) - c_3 z + g_z$$

NONLINEAR GLOBOCOP.

$$\Delta x = -a_1 x + \frac{a_2}{\beta}(y^\beta - (x + C)^\beta) + \frac{a_3}{\beta}(z^\beta - (x + C)^\beta) + g_x$$
$$\Delta y = \frac{b_1}{\beta}(x^\beta - (y + C)^\beta) - b_2 y + \frac{b_3}{\beta}(z^\beta - (y + C)^\beta) + g_y$$
$$\Delta z = \frac{c_1}{\beta}(x^\beta - (z + C)^\beta) + \frac{c_2}{\beta}(y^\beta - (z + C)^\beta) - c_3 z + g_z$$

With increasing $\beta$, which was implicitly set at unity in the linear models above, the volatility of the system grows. Recall that in figure 3.3, with $\beta = 1$, the Richardson model produced monotonic growth to some equilibrium. With all else as in that run, but with $\beta = 1.7$, the dynamics are radically altered. Figure 3.10 gives the result. By any definition, an acute crisis develops. This, strictly speaking, is not chaos, but it is wild.[59] System-wide strategic manic-depression and increasingly explosive oscillations develop.

[59] For a precise definition of chaos, see Devaney (1989, p. 50). On chaotic dynamics in arms race models, see Mayer-Kress (1991).

FIGURE 3.10 Nonlinear Richardson 1

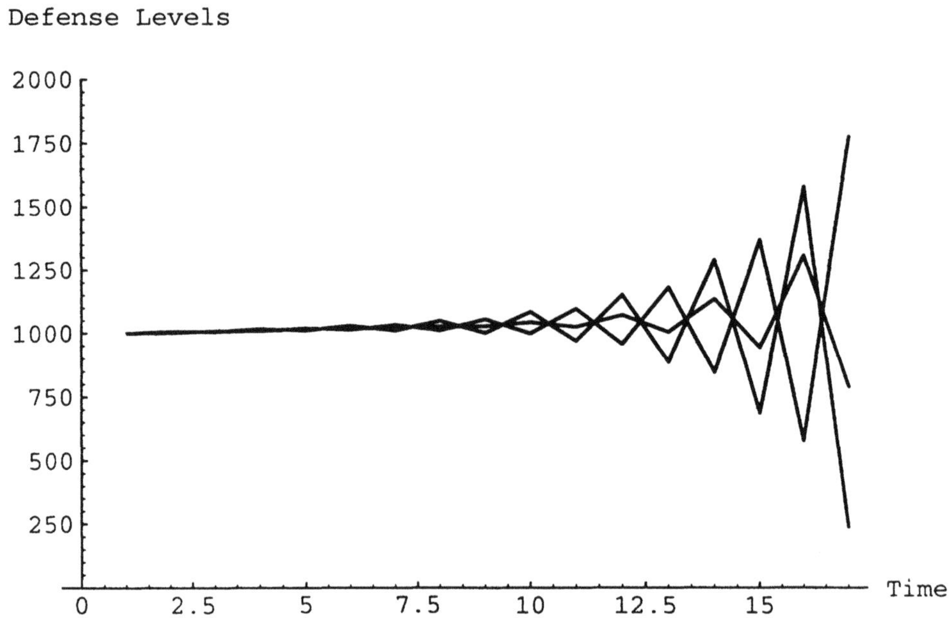


Now, the striking fact is that highly imperfect collective security regimes of various sorts can contain this crisis. First, let us assume maximal connectivity (reciprocal altruism) but diluted strength.

## MAXIMUM CONNECTIVITY WITH LOW STRENGTH

The dilution is captured by $\gamma$ in the collective security equations of figure 3.9 above. If $\gamma = 1$, perfect collective security is restored, with 100 percent of third party forces going instantly to the defender. If $\gamma = 0.5$, half the forces are so allocated, and so forth. Here, of course, we are interested in the weakest forms of collective security that are of interest and surprisingly, with all other numbers set as in the explosive case of figure 3.10, a one-percent solution of collective security ($\gamma = 0.01$) contains the blaze. Crisis management occurs, as shown in figure 3.11a, which displays $x(t), y(t)$, and $z(t)$ superimposed. Figure 3.11b gives a three-dimensional phase portrait of the same dynamics.

FIGURE 3.11 (a) Nonlinear Collective Security 1; (b) Nonlinear Collective Security 1, 3D Phase Portrait

(a)

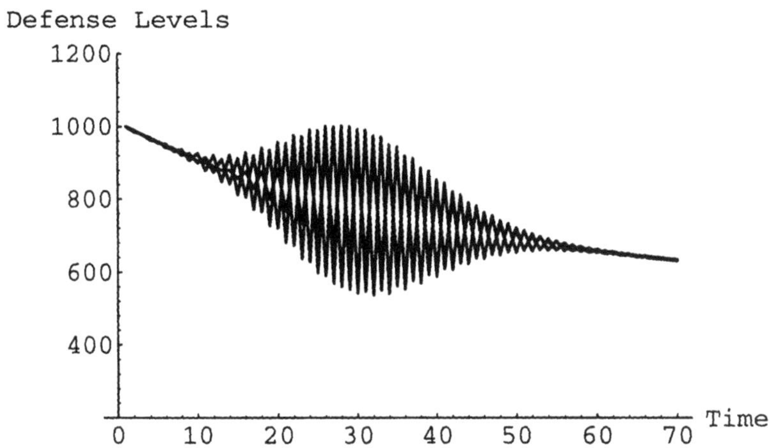


(b)

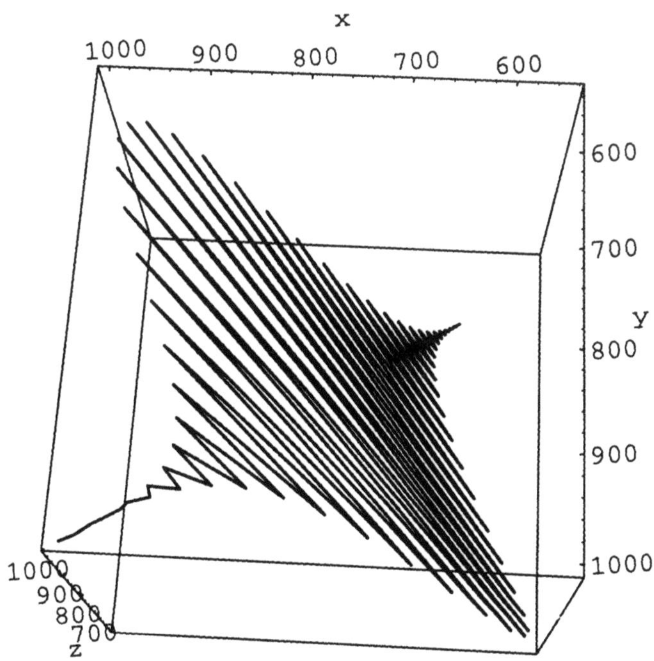


A crisis indeed develops, but it is contained, and the system then calms down to an equilibrium below the initial state. This colorful transient behavior—which we interpret as crisis containment—occurs at the "boundary" between explosive oscillation and monotone disarmament. That is, if we begin with $\gamma = 0$ and the explosive oscillations of figure 3.10, and smoothly increase $\gamma$, we will arrive at a value beyond which all oscillations will be damped away and all trajectories simply decrease monotonically to equilibrium. Some may find even these transient oscillations worrisome. Particularly in the early phase, when the oscillations are increasing in amplitude, the expectation that one's security (i.e., superiority) is transitory and will soon give way to inferiority may generate preemptive (strike while you're ahead) pressures. But even these transient oscillations can be eliminated outright at remarkably low $\gamma$ values. *Ceteris paribus*, $\gamma = .02$—a two-percent commitment—has that effect.[60]

Now, what's going on here; how are such low levels of cooperation producing such dramatic effects? The key is the parameter $\beta$. The higher is $\beta$, the more volatile is the Richardsonian system. But, in collective security, that same $\beta$ is applied to a *sum*—the Richardsonian defender *plus* his collective security ally. The more powerfully are low levels of collective security amplified (by $\beta > 1$), the more will a *little* collective security enhance stability. The greater the underlying volatility (high $\beta$), the greater the stabilizing effect of a given level ($\gamma$) of collective security.[61] One might assume that the more volatile the system is, the less value there will be in a low level of collective security. But, precisely the reverse is true!

By the same token, if we go to a low $\beta$—a less volatile unregulated Richardsonian world—it takes a much stronger transition to collective security to significantly

[60] By the same token, there exists some real $\gamma$ for which the transient phase will have *any* specified duration. I thank Robert Axtell for this thought.

[61] The dependence on $\beta$ of collective security's depressive effect can be analyzed by examining the partial derivative of, say $x$'s growth rate $\Delta x$ with respect to $\gamma$, the degree of collective security, for various $\beta$s. It suffices to consider the term

$$\tau = \frac{a_2}{\beta}(y^\beta - (x + \gamma z)^\beta)$$

from the $\Delta x$ equation of the nonlinear collective security model above. Considering this term only (since the $a_3$ term will behave analogously), compute

$$\frac{\partial \tau}{\partial \gamma} = -a_2 z(x + \gamma z)^{\beta - 1}.$$

Clearly, if $\beta = 1$, then $\gamma$'s depressive effect $(\partial\tau/\partial\gamma)$ varies linearly with $z$. But, at the other extreme of $\beta = 2$, the product of $x$ and $z$, and the *square* of $z$, enter in. Then,

$$\frac{\partial \tau}{\partial \gamma} = -a_2(\gamma z^2 + zx).$$

alter trajectories. To wit, figure 3.12 shows our nonlinear Richardsonian case, but with $\beta = 1.2$ rather than $\beta = 1.7$ from the crisis case.

FIGURE 3.12 Nonlinear Richardson 2

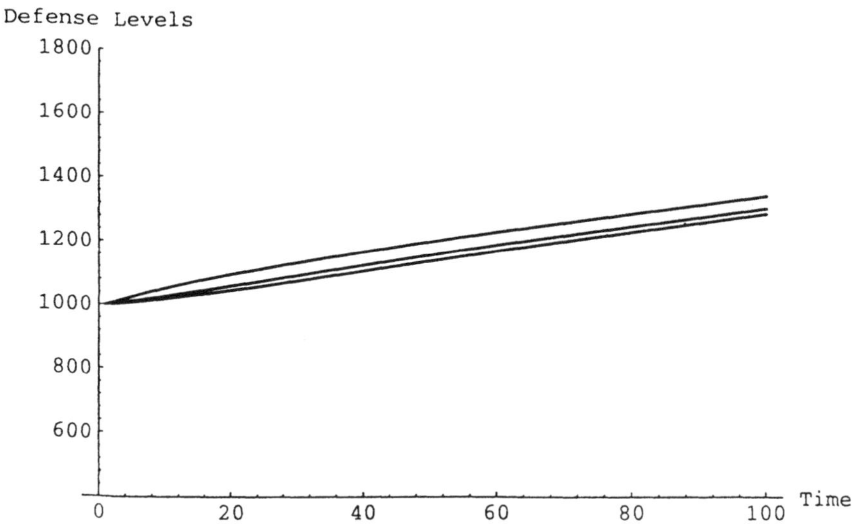


The same level of collective security ($\gamma = 0.01$) that contained the crisis at $\beta = 1.7$ has virtually no effect here, as shown in figure 3.13.

FIGURE 3.13 Nonlinear Collective Security 2 (Low $\gamma$)

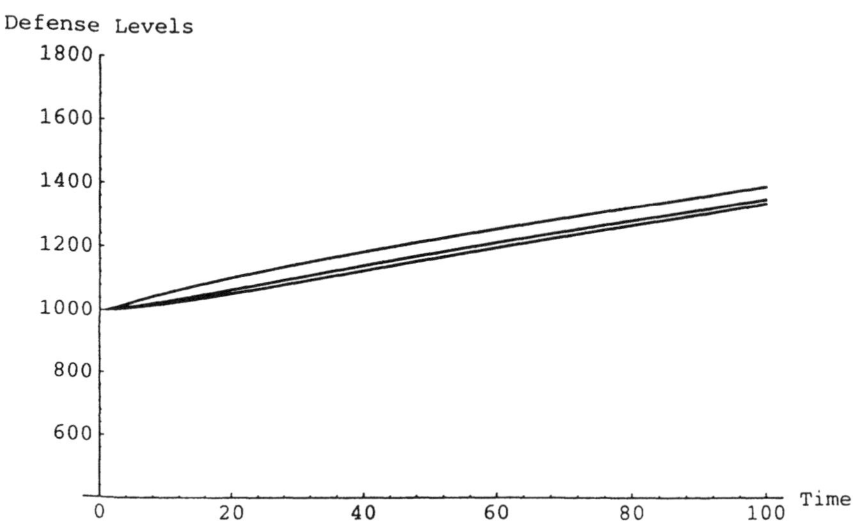

Recall that a one-percent collective security commitment ($\gamma = 0.01$) was sufficient to contain explosive oscillations and produce reductions in the $\beta = 1.7$ case. Here, to merely produce constancy (a weaker requirement than reduction) requires a $\gamma$ of 0.10, a tenfold increase. This case is shown in figure 3.14.

FIGURE 3.14 Nonlinear Collective Security 3 (Higher $\gamma$)

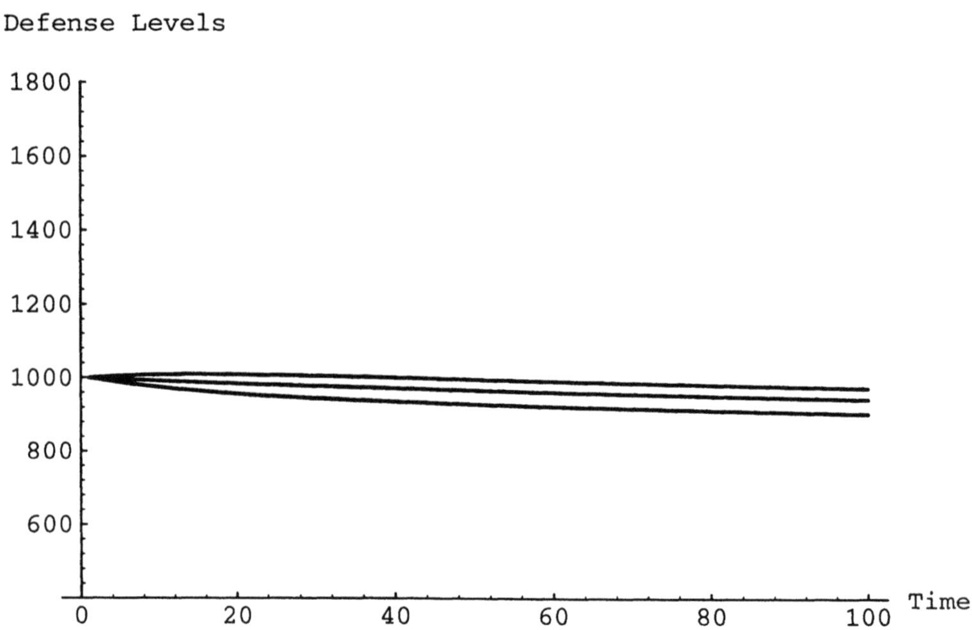


Let us restore the value $\beta = 1.7$ and examine another form of imperfect collective security.

## LOW CONNECTIVITY WITH HIGHER (BUT STILL VERY LOW) STRENGTH

Specifically, let us now relax the assumption of maximal connectivity and posit a low connectivity configuration like the cyclic altruism web of figure 3.7. If we leave the connection strength at $\gamma = 0.01$ as in figure 3.11, crisis containment will not occur; we will see the explosive dynamics of figure 3.10. So, connection strength ($\gamma$) must be raised to compensate for the reduced connectivity. How much greater than 0.01 must the connection strength, $\gamma$, be in order to contain the explosion and force convergence to an equilibrium below initial levels? The value of $\gamma = 0.02$ produces the dynamics shown in figures 3.15a, and 3.15b.

FIGURE 3.15 (a) Cyclic Altruism with $\gamma = 0.02$ (b) Cyclic Altruism with $\gamma = 0.02$, 3D Phase Portrait

(a)

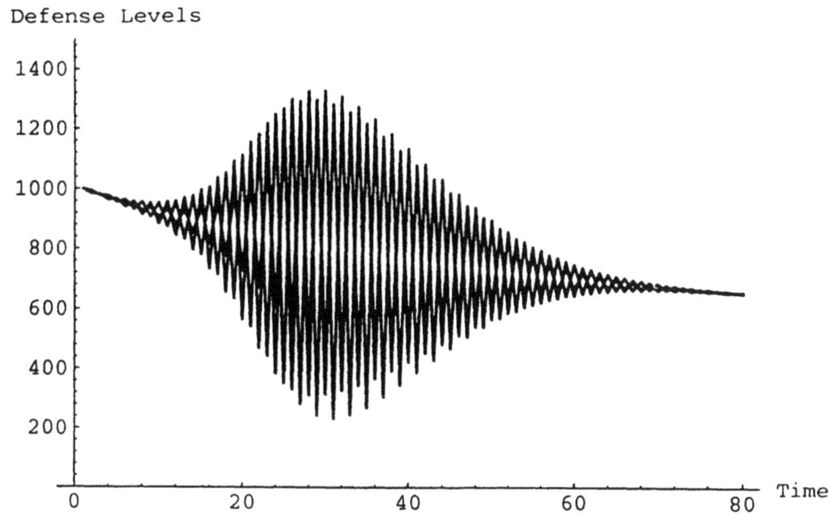


(b)

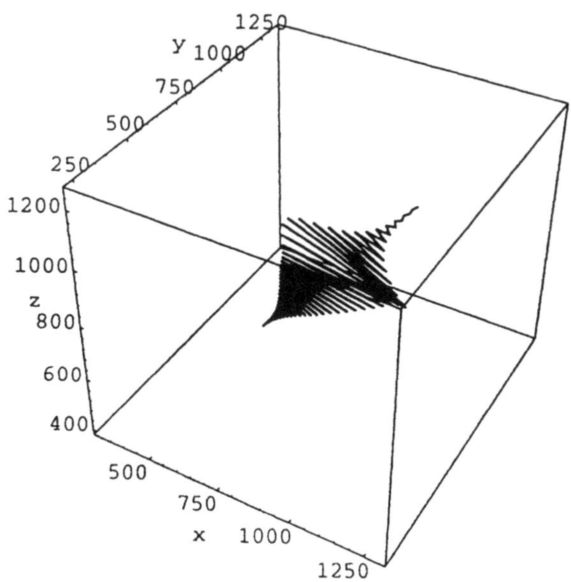


Again, what is striking is that this $\gamma$ is still very low; a two-percent collective security commitment under a sparse connection pattern still suffices.

A comparably low value of $\gamma = 0.025$ also suffices to contain the explosion and force convergence to a low equilibrium under the open connection pattern of figure 3.8. Dynamics in that case are shown in figures 3.16a and b.

FIGURE 3.16 (a) Open Web with $\gamma = 0.025$ (b) Open Web with $\gamma = 0.025$, 3D Phase Portrait

(a)

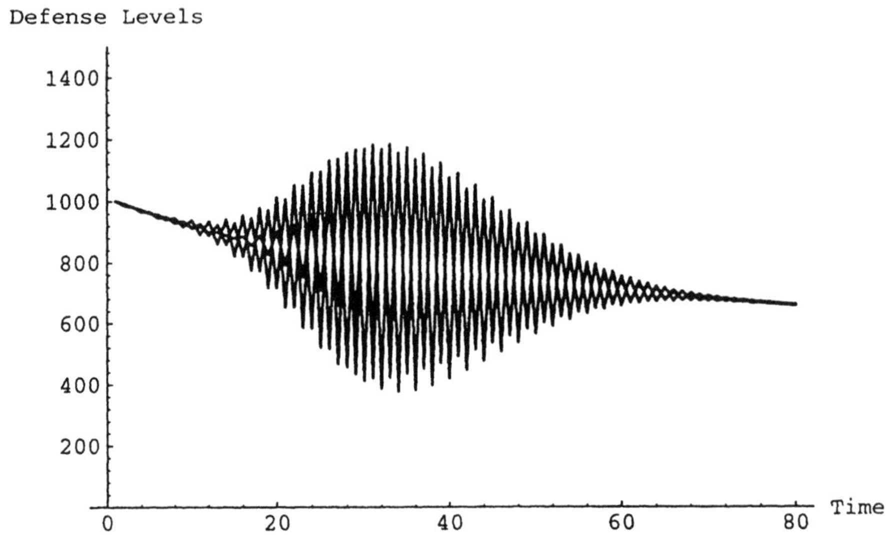


(b)

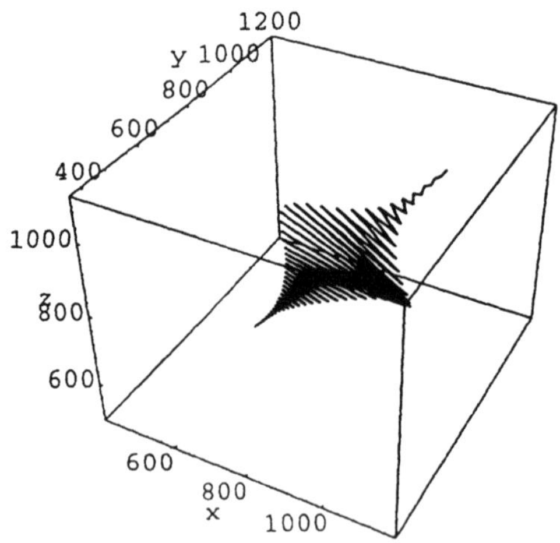


Clearly, collective security can be highly imperfect in a *spectrum of ways* and still exercise a powerfully depressive effect on dynamics.

## GLOBOCOP

Thus far, I have said almost nothing about globocop. Under globocop, when evaluating the threat of external aggression, all actors assume a defensive reinforcement of $C$. In the linear case, globocop is equivalent to Richardson with reduced grievances and is, for that reason, quite uninteresting mathematically. In the nonlinear case, our exponent $\beta$ reenters the picture, amplifying the globocop-augmented defenders substantially for, again, quite modest $C$-values, if $\beta$ is high. At the $\beta$-value of 1.7, under which a one-percent solution of collective security contained the crisis of figure 3.10, a one-percent globocop of $C = 10$ does perfectly well, too. The simulation is given in figures 3.17a and b.

FIGURE 3.17 (a) Nonlinear Globocop; (b) Nonlinear Globocop, 3D Phase Portrait

(a)

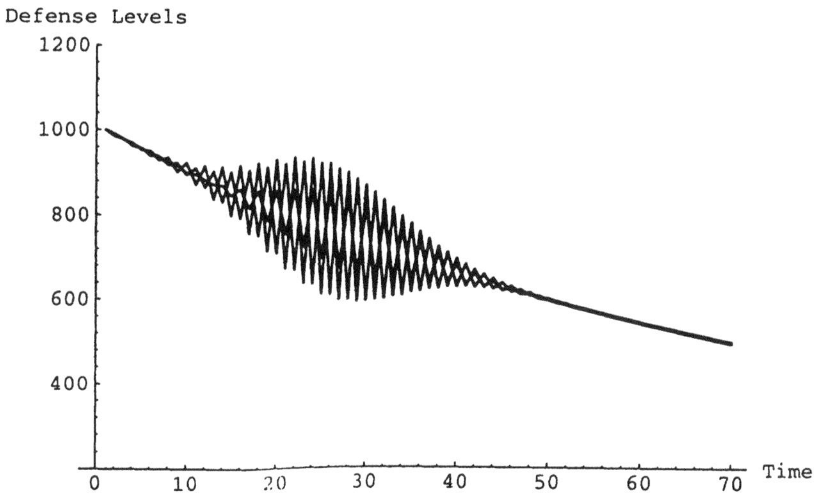


(b)

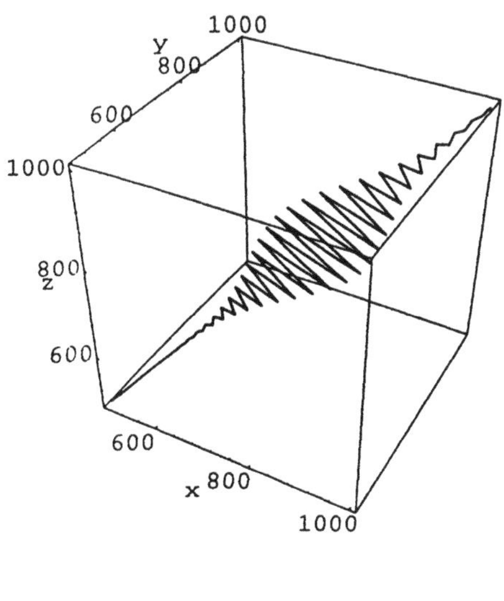

A general nonlinear model that specializes to all the above variants and, notably, to mixes of globocop and collective security is given below.[62]

FIGURE 3.18 General Nonlinear Model

$$\Delta x = -a_1 x + \frac{a_2}{\beta}\left(y^\beta - (x + \gamma z + C)^\beta\right) + \frac{a_3}{\beta}\left(z^\beta - (x + \gamma y + C)^\beta\right) + g_x$$

$$\Delta y = \frac{b_1}{\beta}\left(x^\beta - (y + \gamma z + C)^\beta\right) - b_2 y + \frac{b_3}{\beta}\left(z^\beta - (y + \gamma x + C)^\beta\right) + g_y$$

$$\Delta z = \frac{c_1}{\beta}\left(x^\beta - (z + \gamma y + C)^\beta\right) + \frac{c_2}{\beta}\left(y^\beta - (z + \gamma x + C)^\beta\right) - c_3 z + g_z$$

## STRUCTURAL SENSITIVITY ANALYSIS AND CONCLUSIONS

In the usual sensitivity analysis, one uses a fixed mathematical model and examines the sensitivity of that model's outputs to variations in model inputs. Here, we have fixed inputs (e.g., $x(0) = y(0) = z(0) = 1000$) and examined the sensitivity of output to variations in the model. We started with the linear Richardson model and we perturbed it into a linear collective security variant. Then we constructed a nonlinear Richardson model and perturbed that into a nonlinear collective security variant. In both cases, we found collective security to have a powerfully depressive effect on the competition. I certainly do not claim that either underlying Richardson model is correct in an empirical or predictive sense. If, however, the "collective security effect" is systematically depressive across the entire set of plausible underlying Richardsonian and non-Richardsonian models, it is fair to regard the effect as quite robust. Thus far, of course, the results are only suggestive. But, within the class of models examined here, the following points apply.

With all numbers—propensities to arm in response to perceived imbalances, grievances, and so forth—fixed, modest institutional changes alone can drastically alter dynamics. *Collective security, even in highly imperfect forms, is a powerful depressant. And the depressive (or "stabilizing") effect is the greater the more volatile (in the $\beta$-sense) is the underlying structure.* I have ventured no opinion whatever on the feasibility of implementing collective security anywhere. But, based on this analysis, it would be unfortunate indeed if imperfect collective security arrangements were dismissed simply on the ground that perfect ones may—indeed, probably would—be difficult to institute.[63] Highly imperfect forms of collective security

[62] Kupchan and Kupchan appear to be suggesting some sort of mix between, in effect, globocop (a concert, comprised of "a small group of major powers") and collective security. But it is unclear to me where on the spectrum between these regimes their "concert-based collective security" regime lies. See Kupchan and Kupchan (1995).

[63] Among other things, the purely logistical demands of a working collective security regime could be nontrivial.

may have profound effects and deserve serious study. Depending on the magnitude of $C$, globocop shows similar powerful properties. Indeed, on the basis of this analysis, no especially clear choice between collective security and globocop can be made. Obviously, there may be political arguments pointing to one over the other, but not mathematical ones.[64] It would be interesting to examine whether, within the even wider class of stochastic and game-theoretic arms race models, the same depressive effects of collective security and globocop are suggested.[65]

[64] For instance, collective security raises the classic "free rider" problem. See Olson (1965). On the other hand, globocop has overtones of world government and may be seen as entailing an intolerable sacrifice in national sovereignty.

[65] See Downs and Rocke (1990). Here, utility functions include a term proportional to the gap in arms, so a collective security variant could easily be constructed and examined from this standpoint. The same can be said of the model presented in Forrest and Mayer-Kress (1991, pp. 166–85).

# APPENDIX

TABLE 3.1 Numerical Assumptions for Lecture 3 Figures

| | Figure Number[1] | | | | | | | | | | | |
|---|---|---|---|---|---|---|---|---|---|---|---|---|
| Parameter | 3.2 | 3.3 | 3.4 | 3.5 | 3.10 | 3.11 | 3.12 | 3.13 | 3.14 | 3.15 | 3.16 | 3.17 |
| $a_1$ | 0.003 | | -0.003 | -0.003 | | | | | | | | |
| $a_2$ | 0.0085 | | | | | | | | | | | |
| $a_3$ | 0.0051 | | | | | | | | | | | |
| $b_1$ | 0.0068 | | | | | | | | | | | |
| $b_2$ | 0.002 | | -0.002 | -0.002 | | | | | | | | |
| $b_3$ | 0.0034 | | | | | | | | | | | |
| $c_1$ | 0.0051 | | | | | | | | | | | |
| $c_2$ | 0.0085 | | | | | | | | | | | |
| $c_3$ | 0.001 | | -0.001 | -0.001 | | | | | | | | |
| $\beta$ | 1.0 | | | | 1.7 | 1.7 | 1.2 | 1.2 | 1.2 | 1.7 | 1.7 | 1.7 |
| $\gamma$ | 0.0 | 1.0 | | 1.0 | | 0.01 | | 0.01 | 0.10 | 0.02 | 0.025 | |
| $C$ | 0.0 | | | | | | | | | | | 10.0 |
| $g_x$ | 10.0 | | | | | | | | | | | |
| $g_y$ | 5.0 | | | | | | | | | | | |
| $g_z$ | 3.0 | | | | | | | | | | | |

[1] A blank indicates the same value as in column 1 (figure 3.2).

## LECTURE 4

# Revolutions, Epidemics, and Ecosystems: Some Dynamical Analogies

The preceding lectures have concerned interstate arms racing (mutualism) and war (competition). Let us now turn attention to intrastate processes. This lecture concerns revolutions. The next concerns the spread of drugs. Clinging to our Volterra-like "grand unified theory," the processes of interest in these lectures is the threshold transmission of some "signal" through a population, epidemic-like processes, in short. Epidemics proper are fascinating—and obviously very important—things. You would certainly enjoy William McNeill's wonderful book, *Plagues and Peoples*, which concerns the role of infectious diseases in human history.[66] Since these lectures proceed from the analogy to epidemics, perhaps an introductory word or two on dynamical analogies *per se* is in order.

## ANALOGIES

Any two processes whose mathematical descriptions have the same functional form, and whose state variables and parameters can be put in one-to-one correspondence,

[66] McNeill (1976).

are said to be *dynamical analogies*. It is a startling fact that a huge variety of seemingly unrelated processes are analogous in this sense. For example, the same equation that describes a damped harmonic oscillator, such as a pendulum with friction, also describes an oscillating electric circuit:

> "all that is required is to relabel the state variables and parameters involved. Thus, the state variable representing the displacement of the mechanical system becomes the electrical charge of the electrical system; the velocity becomes the current; the mass of the particle becomes the inductance, mechanical force becomes EMF, etc. With similar reinterpretations, the same dynamical equation can be regarded as describing rotational systems, acoustic systems, hydraulic systems, and so on" (Rosen, 1970, p. 54).[67]

Another example is the analogy between electrostatic attraction under Coulomb's Law and gravitational attraction under Newton's Law. The magnitude of each force is proportional to the product of the two charges/masses, inversely proportional to the square of the distance separating them, and directed along the line joining them. As another instance, Kelvin's circulation theorem in fluid mechanics is identical in its mathematical form to Faraday's Law in electrodynamics. Both relate, via Stokes' Theorem, the flux of a vector field to the circulation (or current) in a boundary such as a conducting loop.[68] Countless further examples could be provided. The physical diversity of diffusive processes satisfying the "heat" equation, or oscillatory processes satisfying the "wave" equation, is virtually boundless.

But dynamical analogies are more than beautiful testaments to the unifying power of mathematics: they are *useful*. In particular, "Analogies are useful for analysis in unexplored fields. By means of analogies an unfamiliar system may be compared with one that is better known. The relations and actions are more easily visualized, the mathematics more readily applied and the analytical solutions more readily obtained in the familiar system."[69]

Analogy in this sense has played a powerful role in the development of science, engineering, and also social science, a notable example of the latter being Samuelson's application to economics of classical maximum principles of physics. In one colorful discussion, for instance, he argues that "if you look at the monopolistic firm as an example of a maximum system, you can connect up its structural relations with those that prevail for an entropy-maximizing thermodynamic system. Pressure and volume, and for that matter absolute temperature and entropy, have to each other the same conjugate or dualist relation that the wage rate has to labor or the land rent has to acres of land." Samuelson provides an elegant diagram that, in

[67] EMF is electromotive force.

[68] See, for example, Marsden and Tromba (1976, p. 338).

[69] Olson (1958, p. iv).

his words, does "double duty, depicting the economic relationships as well as the thermodynamic ones."[70]

Murray Gell-Mann has written on the application of nonlinear dynamics to various systems, including social systems. In his words,

> "Many of these applications are highly speculative. Furthermore, much of the theoretical work is still at the level of 'mathematical metaphor.' But, I think this situation should cause us to respond with enthusiasm to the challenge of trying to turn these metaphorical connections into real scientific explanations" (Gell-Mann, 1988, p. 4).

It is in this highly speculative, metaphorical spirit that I proceed in the next two lectures. This essay examines the analogy between epidemics (for which a well-developed mathematical theory exists) and processes of explosive social change, such as revolutions (for which no comparable body of mathematical theory exists). Are revolutions "like" epidemics? More precisely, is it useful to think of these processes as analogous? Connections to predator-prey systems are also explored and the spatio-temporal generalizations of these revolution/epidemic/predator-prey models—reaction-diffusion equations—are examined. The realm of reaction-diffusion equations is a natural one to explore. Such equations are central to the mathematical theory of pattern formation, and it is the evolution, propagation, and stability of social patterns that is, ultimately, our concern. We begin with the simple analogy between revolutions and infectious diseases.[71]

## REVOLUTIONS AS EPIDEMICS

The particular aims of revolutionary action, of course, vary widely from case to case. In one instance, the revolutionaries' goal may be the overthrow of monarchy; in another, it may be the installation of theocracy; in yet a third, it may be the establishment of democracy. Given this enormous variation in objectives, the thought that there might be an underlying structure common to all revolutions is an intriguing one. By a "common structure," I of course mean a mathematical model whose dynamics—at least at some crude level—are mimicked by revolutionary processes in general, regardless of their political "substance," as it were—regardless, that is,

[70] Samuelson (1972, pp. 8–9).

[71] While the connections between revolutions, epidemics, and ecological systems presented here have not, to my knowledge, been presented elsewhere, the thought that the spread of ideas might be analogous to the spread of disease has been explored. A small literature sprang up in the 1960's, following the publication in 1957 of the seminal work, Bailey (1957). For a good overview with references, see Dietz (1967). See also Rappaport (1974, pp. 47–59).

of what the revolution "is about." In considering this notion, I take, as one tantalizing point of departure, the mathematical theory of epidemics; these processes exhibit common dynamical structures despite obvious differences among communicable diseases. Measles, mumps, and smallpox are clearly different diseases; yet their *propagational dynamics* may be indistinguishable from a mathematical standpoint.[72] Although the points of correspondence between epidemics and revolutions will be quickly evident, it will prove useful to delay specific analogizing until a simple epidemic model is presented.

## A BASIC EPIDEMIC MODEL

The epidemiologist's problem, as Paul Waltman puts it, "is to describe the spread of an infection within a population. As a canonical example one thinks of a small group of individuals who have a communicable infection being inserted into a large population of individuals capable of 'catching' the disease. Then an attempt is made to describe the spread of the infection in the larger group."[73] In the simple model first developed by Kermack and McKendrick,[74] the population is assumed to be constant and divided into three disjoint classes:

$S(t)$: the susceptible class comprised of individuals who, though not infective, are capable of becoming infective;

$I(t)$: the infective class, comprised of individuals capable of transmitting the disease to others; and

$R(t)$: the removed class, consisting of those who have had the disease and are dead, or who have recovered and are permanently immune, or are isolated until recovery and permanent immunity occur.

The following rules are assumed to govern the spread of the disease:

(i) The population is constant over the time interval of interest. Births, deaths from causes other than the disease in question, immigration, and emigration are all ignored.

(ii) The rate of change of the susceptible class is proportional to the product of the number of susceptibles $S(t)$ and the number of infectives $I(t)$.

(iii) Individuals are removed from the infectious class at a rate proportional to $I(t)$.

Rule (i) is a straightforward simplifying assumption whose relaxation is discussed below. Rule (ii) represents the assumption that the transfer of individuals from the

[72] See Hethcote (1976, p. 336). See also Hethcote (1989, pp. 119–44).

[73] Waltman (1974, p. 1).

[74] Kermack and McKendrick (1927). See Murray (1989, pp. 611–18).

susceptible class into the infectious pool proceeds at a rate proportional to the number of contacts between infectives and susceptibles. That the contact rate should be proportional to the product of the class sizes $I$ and $S$ implies uniform mixing of the two groups and instantaneous contraction of the disease upon exposure (latency and incubation periods are both zero). In effect, the law of mass action is assumed to apply. As Waltman notes, "This is reasonable if the population consists of students in a school whose changing classes, attending athletic events, etc. mix the population." Importantly for our purposes, he continues, "It would not be true in an environment where socio-economic factors have a major influence on contacts."[75] Finally, rule (iii) implies that all infectives have the same probability of removal (recovery, death, or isolation). The model does not account for the length of time an individual has been infective.

Accepting these definitions and rules, and treating the overall population as a continuum, the flow of individuals from the susceptible to the infective to the removed class is described by the following system of nonlinear differential equations:[76]

$$\begin{aligned}\frac{dS}{dt} &= -rSI\,,\\ \frac{dI}{dt} &= rSI - \gamma I\,,\\ \frac{dR}{dt} &= \gamma I\,,\end{aligned} \tag{4.1}$$

with initial conditions $S(0) = S_0 > 0, I(0) = I_0 > 0$ and $R(0) = 0$.

The constants $r$ and $\gamma$ are called the infection rate and the removal rate, and $\rho = \gamma/r$ is termed the relative removal rate.

## THE THRESHOLD CONDITION

Now, under what conditions will an epidemic occur in this model? To say that an epidemic occurs is to say that the infectious class grows or, equivalently, that $dI/dt > 0$, which from (4.1) implies that $rSI - \gamma I > 0$ or, simply, that

$$S > \frac{\gamma}{r} = \rho\,. \tag{4.2}$$

[75]Waltman (1974, p. 2).

[76]Because the flow is from susceptible ($S$) to infective ($I$) to removed ($R$), this is termed an $SIR$ model. If the infectious phase is followed, not by removal (e.g., immunity), but by reentry into the susceptible pool, an $SIS$ model would be called for. "In general, $SIR$ models are appropriate for viral agent diseases such as measles, mumps, and smallpox, while $SIS$ models are appropriate for some bacterial agent diseases such as meningitis, plague, and venereal diseases, and for protozoan agent diseases such as malaria and sleeping sickness." Hethcote (1976, p. 336). See also Hethcote (1989). The cornerstone of the mathematical epidemiology literature remains Bailey (1957). See also Bailey (1975). A comprehensive contemporary text is Anderson and May (1991).

This is a basic result.[77] For an epidemic to occur, the number of susceptibles must exceed the threshold level $\rho$—the relative removal rate defined above.

## POLITICAL INTERPRETATION

The basic analogy to revolutionary dynamics is direct. The infection, or disease, is of course the revolutionary idea. The infectives $I(t)$ are individuals who are actively engaged in articulating the revolutionary vision and winning over ("infecting") the susceptible class $S(t)$, comprised of those who are receptive to the revolutionary idea, but who are not infective (not actively engaged in transmitting the disease to others). Removal is most naturally interpreted as the political imprisonment of infectives—$R(t)$ is the "Gulag" population, the set of unfortunate revolutionaries who have been captured and isolated from the susceptible population.[78]

Many familiar tactics of totalitarian rule can be seen as measures to minimize $r$ (the effective contact rate between infectives and susceptibles) or maximize $\gamma$ (the rate of political removal). Press censorship and the systematic inculcation of counterrevolutionary beliefs reduce $r$, while increases in the rate of domestic spying (to identify infectives) and of imprisonment without trial increase $\gamma$.

Symmetrically, familiar revolutionary tactics—such as the publication of underground literature, or "samizdat"—seek to increase $r$. Similarly, Mao's directive that revolutionaries must "swim like fish in the sea," making themselves indistinguishable (to authorities) from the surrounding susceptible population, is intended to reduce $\gamma$.

## GORBACHEV, DeTOQUEVILLE, AND SENSITIVITY TO INITIAL CONDITIONS

Interpreting relation (4.2) somewhat differently, if the number of susceptibles, $S_0$, is in fact quite close to $\rho$, then even modest reductions (voluntary or not) in central authority can push society over the epidemic threshold, producing an explosive overthrow of the existing social order. To take the example of Gorbachev, the policy of Glasnost obviously produced a sharp increase in $r$, while the relaxation of political repression (e.g., the weakening of the KGB, the release of prominent political prisoners, the dismantling of Stalin's Gulag system) constituted a reduction in $\gamma$. Combined, these measures evidently depressed $\rho$ to a level below $S_0$, and the "revolutions of 1989" unfolded. Perhaps DeToqueville intuited relation (4.2), describing

[77] Obviously, the system (4.1) has a great many further mathematical properties of interest. For a discussion, see Braun (1983, pp. 456–73).

[78] In this discussion, we ignore executions.

this sensitivity to initial conditions, when he remarked that "liberalization is the most difficult of political arts."

## TRAVELING WAVES

In the discussion thus far, the *spatial* dimension has only been implicit. In fact, epidemics spread across geographical areas over time. And one generally thinks of revolutions spreading as well. Specifically, we often invoke the terminology of *waves*. Recently, we saw "a wave of democratic revolutions" sweep across Eastern Europe. Perhaps this sort of language seems natural for a reason: if one generalizes model (4.1) to explicitly include the *spatial diffusion* of infectives, traveling waves do indeed emerge. And this process, of course, has a political interpretation.

The one-dimensional spatio-temporal generalization of (4.1) is:

$$\begin{aligned} \frac{\partial S}{\partial t} &= -rIS\,, \\ \frac{\partial I}{\partial t} &= (rIS - \gamma I) + D\frac{\partial^2 I}{\partial x^2}\,. \end{aligned} \tag{4.3}$$

An infective spatial diffusion term, $D\partial^2 I/\partial x^2$, has been introduced into the second equation, which bears some resemblance to the classical heat equation, $I_t = DI_{xx}$, where $D$ is the thermal—or, in this case, the political—"diffusivity" of the medium. The presence of the parenthesized term makes the equation a so-called reaction-diffusion relation.

Now, as set forth in lecture 6, one posits traveling wave solutions to (4.3) of the form

$$S(x,t) = S(z), \quad I(x,t) = I(z), \quad z = x - ct\,, \tag{4.4}$$

where $c$ is the wave speed. The boundary conditions $S(\infty) = 1, S(-\infty) = 0, I(\infty) = I(-\infty) = 0$ must also be met. Bypassing mathematical specifics that are well presented elsewhere,[79] the basic conclusions are, first, that no epidemic wave propagates if $S_0 < \gamma/r$. This, of course, is the basic threshold condition from model (4.1). What is new, however, is that if that threshold level of susceptibility is exceeded, an epidemic/revolutionary *wavefront* propagates. And its speed of propagation, $c$, is given by

$$c = 2[D(rS_0 - \gamma)]^{1/2}\,. \tag{4.5}$$

Basic counterrevolutionary tactics aim not only to minimize $r$ (the rate at which contact produces a transmission) and maximize $\gamma$ (the removal rate), but to minimize $D$ as well. Physical curfews, restrictions on free assembly, internal

[79]See Murray (1989, pp. 661–63), Britton (1986, pp. 61–71), and the discussion in lecture 6 of this volume.

passport requirements, apartheid in all its forms, are means to "patchify," and limit the "political diffusivity" of, the social medium. A high density of internal police reduces the diffusion of revolutionaries just as a high density of wet insects retards the spread of a brush fire.[80] Indeed, the most obvious "firebreaks," or $D$-minimizers, are the borders between countries, which physically enforce certain of the ideological "patches" dividing humanity.[81]

## VITAL DYNAMICS AND THE EVOLUTION OF DISCONTENT

Thus far in the discussion, the total population has been assumed constant. Processes of profound social change may unfold over periods in which birth and death—what epidemiologists call "vital dynamics"—play a role. The introduction of these factors expands the range of possible revolutionary/epidemic trajectories. Indeed, the introduction of some rudimentary vital dynamics connects our discussion, perhaps surprisingly, to the field of mathematical ecology.

For expository ease, let us recall the basic epidemic model (4.1). Keeping matters simple, now introduce a Malthusian birth rate into the susceptible population. If $\mu > 0$ is the birth rate, we obtain the system

$$\begin{aligned} \frac{dS}{dt} &= -rSI + \mu S\,, \\ \frac{dI}{dt} &= rSI - \gamma I\,. \end{aligned} \tag{4.6}$$

Students of mathematical ecology will recognize this as precisely the Lotka-Volterra predator-prey model. The prey (susceptibles) would increase exponentially if not for the predators (infectives), who would die off exponentially without their "food source," the prey. Once a birth term is introduced, the infectives and susceptibles may be seen as forming an ecosystem.

As shown in figure 4.1, the orbits of (4.6) are closed curves in the $SI$ phase plane; solutions are undamped oscillations.[82]

[80] See Murray (1989, pp. 375–76).

[81] A basic strategy to control the spread of rabies in foxes, for example, is to thin the susceptible fox population to a level below the critical threshold density across some swath lying in front of the advancing epidemic wave; to create a firebreak, as it were. For an analytic approximation for the minimum width of a rabies control break, as such barriers are known, see Murray (1989, pp. 681–89).

[82] A damped variants results if $\mu S$ is replaced simply by $\mu$. See Bailey (1975, pp. 135–37).

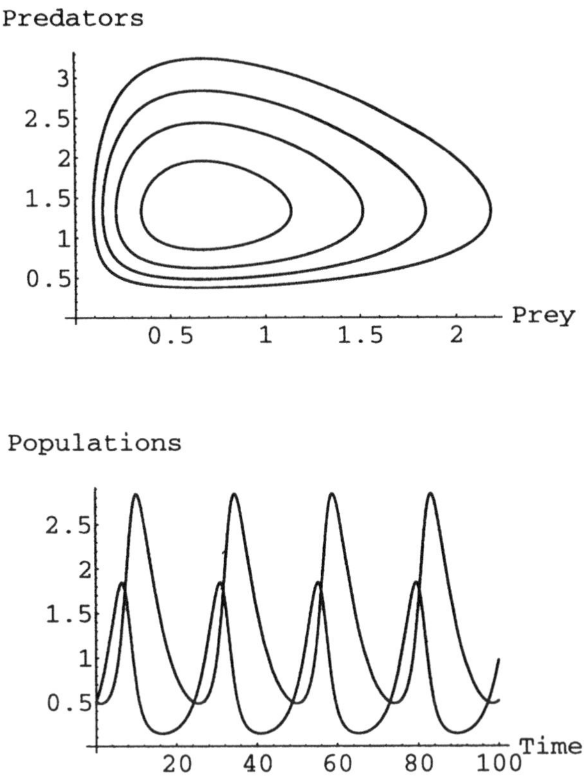


FIGURE 4.1 Predator-Prey Cycles and a Periodic Solution

The political interpretation is straightforward: extending model (4.1) to include rudimentary vital dynamics ("net births"), as in (4.6), society may experience periodic cycles of revolutionary discontent (the infection) analogous to cycles characteristic of basic model ecosystems, and recurrent epidemics.[83] Interestingly, Goodwin's model of the class struggle between workers and capitalists is of exactly the same form.[84]

In the physical sciences, it is often of great interest to identify entities that are conserved by the dynamical system—functions that are constant along system trajectories. As discussed in lecture 6, such functions are termed Hamiltonians of

[83] On cyclical behavior in epidemic and ecological models specifically, the interested reader might see Hethcote and Levin (1989, pp. 192–211) and Huang and Merrill (1989).

[84] Goodwin (1967). The model is clearly discussed in Lorenz (1989).

the system. Interestingly, this model admits a Hamiltonian formulation. Under the transformation $p = \ln S$, $q = \ln I$, system (4.6) becomes[85]

$$\begin{aligned} \frac{dp}{dt} &= \mu - re^q, \\ \frac{dq}{dt} &= re^p - \gamma. \end{aligned} \tag{4.7}$$

Separating variables and integrating the equation

$$\frac{dp}{dq} = \frac{\mu - re^q}{re^p - \gamma},$$

we obtain

$$re^p - \gamma p + re^q - \mu q = c,$$

where $c$ is a constant determined by initial conditions. The left-hand side, constant along trajectories, defines a Hamiltonian:

$$H(p, q) = re^p - \gamma p + re^q - \mu q. \tag{4.8}$$

It is easy to verify that (4.7) is, equivalently, Hamilton's equations:

$$\begin{aligned} \dot{p} &= \frac{-\partial H}{\partial q}, \\ \dot{q} &= \frac{\partial H}{\partial p}. \end{aligned} \tag{4.9}$$

A social interpretation of this energy-like Hamiltonian—a constant of revolutionary/counter-revolutionary motion—would be intriguing.

This purely oscillatory—neutrally stable—behavior of system (4.6) breaks down when the assumption of Malthusian growth is abandoned. Indeed, the model is structurally unstable in the language of lecture 6. Assuming some carrying capacity for the susceptibles, a more plausible system would be

$$\begin{aligned} \frac{dS}{dt} &= -rSI + \mu S\left(1 - \frac{S}{K}\right), \\ \frac{dI}{dt} &= rSI - \gamma I, \end{aligned} \tag{4.10}$$

where $K$ is the carrying capacity.

[85] See Verhulst (1990, pp. 21–22). For further generalization, and an elegant discussion of Hamiltonians in this context, see Samuelson (1971).

Once more, the introduction of spatial diffusion terms renders (4.10) into a reaction-diffusion system. In one spatial dimension,

$$\begin{aligned}\frac{\partial S}{\partial t} &= -rSI + \mu S\left(1 - \frac{S}{K}\right) + D_1 \frac{\partial^2 S}{\partial x^2}\,,\\ \frac{\partial I}{\partial t} &= rSI - \gamma I + D_2 \frac{\partial^2 I}{\partial x^2}\,.\end{aligned} \tag{4.11}$$

And, as in the case of the generalized epidemic model (4.3), traveling wavefront solutions exist. Assuming for consistency's sake that it is again the infectives that diffuse, let $D_1/D_2 = 0$. Then, under the usual boundary conditions,[86] a minimum wavefront velocity, $c_{\min}$, is given by

$$c_{\min} = \left[\frac{4rK}{\mu}\left(1 - \frac{\gamma}{rK}\right)\right]^{1/2}, \quad \frac{\gamma}{rK} < 1\,. \tag{4.12}$$

An interesting difference between (4.11) and (4.3), however, is that this wavefront solution can be approached in an oscillatory or monotonic fashion depending on parameter values.[87]

## THE REMOVERS

The models thus far presented place no upper bound on $R(t)$; there is an unlimited removal capacity. One refinement is the imposition of some upper bound on $R(t)$. Second, the counter-revolutionary class (the elite and its gendarmerie), or "public health authority," is not explicitly represented. Clearly, since the power elite is doing the removing, it deserves a place in the model. Let us denote this group by $E(t)$ and, for simplicity, assume it to be constant at some initial level $E_0$. Then, if $G$ denotes the upper bound on removals, a slightly refined version of our first (Kermack-McKendrick) model might take the following form:

$$\begin{aligned}\frac{dS}{dt} &= -rSI\,,\\ \frac{dI}{dt} &= rSI - \alpha E_0 I\,,\\ \frac{dR}{dt} &= \alpha E_0 I\,.\end{aligned} \tag{4.13}$$

[86] Murray (1989, p. 316).
[87] Murray (1989, p. 318).

The basic removal (repression) rate parameter $\gamma$ has been supplanted by $\alpha E_0$, where $\alpha > 0$ if $R(t) < G$ and $\alpha = 0$ otherwise. In other words, $\alpha$ equals a positive constant so long as there is "room in the Gulag," but equals zero once the Gulag is full (or, more generally, once the removal capacity of the State is reached).

We saw that, under the model (4.1), with no upper bound on removals, the infection ultimately dies out. At the other extreme, if $G = 0$ (no removal capacity), the transfer of susceptibles into the infective class—that is, the spread of the infection—is in effect governed by the system

$$\begin{aligned} \frac{dS}{dt} &= -rSI\,, \\ \frac{dI}{dt} &= rSI\,. \end{aligned} \tag{4.14}$$

Since the total population $P$ and the elite $E$ are constants, we have $S(t) + I(t) + E_0 = P_0$, or $S(t) = [P_0 - E_0] - I(t)$, and hence can write

$$\begin{aligned} \frac{dI}{dt} &= r([P_0 - E_0] - I(t))I(t) \\ &= cI - rI^2\,, \end{aligned} \tag{4.15}$$

a Bernoulli equation whose solution,

$$I(t) = \frac{cI_0}{rI_0 + (c - rI_0)e^{-ct}}\,, \tag{4.16}$$

is precisely the Logistic *function*. This same function, interestingly, has been found to govern the diffusion of certain technological innovations—"technology epidemics."[88] The closely related discrete Logistic *map* is, of course, the prototypical chaotic dynamical system. And, in fact, the question whether epidemics exhibit chaotic behavior is under active study.[89] Such connections, it seems to me, are potentially quite interesting.

## IMPERMANENT REMOVAL

Thus far we have assumed any removals to be permanent; the Kermack-McKendrick flow is from $S$ to $I$ to $R$. In fact, even when the state's removal capacity is unbounded (a reasonable assumption in most practical cases), removal need not be permanent. Removed individuals may eventually reenter the susceptible pool—as when we reemerge from a stay in the hospital, or the prison, as the case may be. Models capturing this are, for obvious reasons, termed $SIRS$ models.[90] Extending

[88] See Mansfield (1961). See also Cavalli-Sforza and Feldman (1981).

[89] See, for example, Olsen and Schaffer (1990). For a rigorous mathematical definition of the much-abused term "chaos," see, for example, Devaney (1989).

[90] In principle, one could return from prison to the infective, rather than susceptible, pool, producing an $SIRI$ model.

Kermack-McKendrick, we obtain the system

$$\frac{dS}{dt} = -\beta SI + \nu R\,, \tag{4.17}$$

$$\frac{dI}{dt} = \beta SI - \gamma I\,, \tag{4.18}$$

$$\frac{dR}{dt} = \gamma I - \nu R\,, \tag{4.19}$$

where $\gamma = \alpha E_0$ from the previous (political) interpretation.[91] We are most interested in the positive equilibrium. Clearly, from (4.18), $\bar{S} = \gamma/\beta$, and we can write this equilibrium

$$(\bar{S}, \bar{I}, \bar{R}) = \left(\bar{S}, \nu\frac{N-\bar{S}}{\nu+\gamma}, \frac{\gamma\bar{I}}{\nu}\right) . \tag{4.20}$$

Since, in this model, population is constant at $N$, we may write $R = N - S - I$, so that the system, in $SI$-space, becomes

$$\frac{dS}{dt} = -\beta SI + \nu(N - S - I) = f_1(S, I)\,, \tag{4.21}$$

$$\frac{dI}{dt} = \beta SI - \gamma I = f_2(S, I)\,. \tag{4.22}$$

Denoting by $F$ the vector field $(f_1, f_2)$, the Jacobian at any equilibrium $\bar{x} = (\bar{S}, \bar{I})$ is

$$DF(\bar{x}) = \begin{pmatrix} -(\beta\bar{I} + \nu) & -(\beta\bar{S} + \nu) \\ \beta\bar{I} & \beta\bar{S} - \gamma \end{pmatrix} .$$

At the positive (or interior) equilibrium of interest, $\bar{S} = \gamma/\beta$, so that $\beta\bar{S} - \gamma = 0$ and we have

$$DF(\bar{x}) = \begin{pmatrix} -(\beta\bar{I} + \nu) & -(\nu + \gamma) \\ \beta\bar{I} & 0 \end{pmatrix} .$$

Clearly, the trace is negative and the determinant is positive; so, as reviewed in lecture 6, this equilibrium is *stable*. With $T$ the trace and $D$ the determinant of $DF(\bar{x})$, its eigenvalues are:

$$\lambda_1, \lambda_2 = \frac{1}{2}\left(T \pm \sqrt{T^2 - 4D}\right) .$$

The real parts are negative, but we may have nonzero imaginary parts, and spiral convergence to the equilibrium, as shown in figure 4.2.

[91]This analysis parallels Edelstein-Keshet (1988, pp. 245–49).

FIGURE 4.2 Spiral Sink.

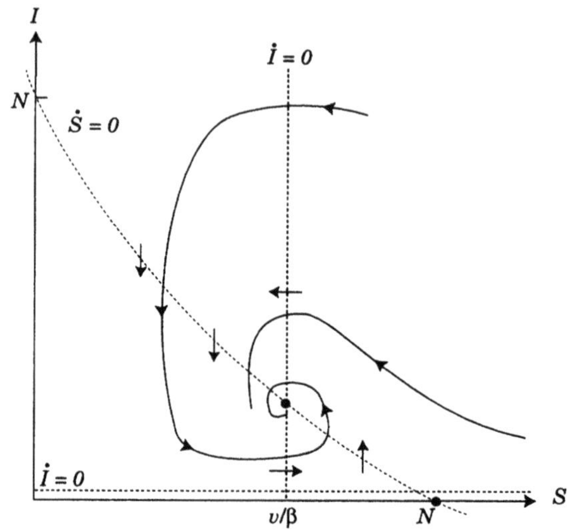


Source: Based on Edelstein-Keshet (1988, p. 248).

The political interpretation of spiral convergence would be periodic outbursts that are repressed to a steady level of endemic positive discontent.

Now suppose—as a *gedanken* experiment—that you are a Machiavellian social engineer and want to ensure that revolutionary ideologies pose no threat to the established order. One instrument is indoctrination. The parallel to inoculation may be instructive.

## HERD IMMUNITY AND DECENTRALIZED TOTALITARIANISM

Staying with the same model, we see from (4.20) that for $\bar{I}$—the endemic level—to be positive we must have $N - \bar{S} > 0$ which is to say $N\beta/\gamma > 1$. With May, we define $R_0 = N\beta/\gamma$ to be the intrinsic reproductive rate of the disease.[92] "$R_0$ represents the average number of secondary infections caused by introducing a single infected individual into a host population of susceptibles."[93] The epidemic threshold condition is then simply $R_0 > 1$.

[92] May (1983). $R_0$ is also termed the "reproductive number."

[93] Edelstein-Keshet (1988, p. 247).

Suppose, however, that we can vaccinate some fraction, $P$, of the population (still constant at $N$). The fraction immunized should be big enough that the unimmunized fraction $(1-P)N$ is *below* the threshold. We will then have achieved "herd immunity." Proceeding, we require

$$[(1-P)N]\frac{\beta}{\gamma} < 1\,, \text{ or}$$
$$(1-P)R_0 < 1\,,$$

from which the required inoculation level is given by

$$P > 1 - \frac{1}{R_0}\,. \tag{4.23}$$

For various diseases, we have table 4.1.

TABLE 4.1 Immunization Levels $P$ Required for Herd Immunity: Various Diseases.

| Infection | Location and Time | $R_0$ | Approximate Value of $P$(%) |
|---|---|---|---|
| Smallpox | Developing countries, before global campaign | 3–5 | 70–80 |
| Measles | England and Wales, 1956–68; | 13 | 92 |
| | U.S., various places, 1910–30 | 12-13 | 92 |
| Whooping cough | England and Wales 1942–50; | 17 | 94 |
| | Maryland, U.S., 1908–17 | 13 | 92 |
| German measles | England and Wales, 1979; | 6 | 83 |
| | West Germany | 7 | 86 |
| Chicken pox | U.S., various places, 1913–21 and 1943 | 9–10 | 90 |
| Diphtheria | U.S., various places, 1910–47 | 4–6 | ~80 |
| Scarlet fever | U.S., various places, 1910–20 | 5–7 | ~80 |
| Mumps | U.S., various places, 1912–16 and 1943 | 4–7 | ~80 |
| Poliomyelitis | Holland, 1960: U.S., 1955 | 6 | 83 |

Source: Edelstein-Keshet (1988, p. 255).

If we now recover our political analogues, we have the required level of social indoctrination given by

$$P > 1 - \frac{\alpha E_0}{N\beta}\,. \tag{4.24}$$

An increase in domestic police ($E_0$) or in their effectiveness ($\alpha$) will allow the elite to do less indoctrination, while an increase in $\beta$ (the infectiveness of revolutionary ideas) will require increased propagandistic effort, all of which makes a certain amount of sense. A somewhat darker reading is invited by the rearrangement

$$\frac{\alpha E_0}{N(1-P)} > \beta\,. \tag{4.25}$$

If indoctrination is very high, even the most compelling revolutionary critique will pose no threat to the established order because too few will listen. Once herd immunity is achieved, the system is really on autopilot; a type of *decentralized totalitarianism* is possible.

## BIFURCATION

Regarding another crucial variable, repression, there are many subtleties that one might wish to capture. For example, in some situations, repression has a deterrent, or "chilling," effect, while in other situations, it simply inflames hostility toward the regime. Indeed, revolutionaries have been known to provoke repression with precisely this inflammatory aim. Moreover, a given population may flip from one "response mode" to the other in the wake of particular incidents. With these thoughts in mind, consider the following extension, which was proposed by Jean-Pierre Langlois.

$$\frac{dS}{dt} = -rSI + \nu R + \delta RI\,, \tag{4.26}$$

$$\frac{dI}{dt} = rSI - \alpha E_0 I - \delta RI\,, \tag{4.27}$$

$$\frac{dR}{dt} = \alpha E_0 I - \nu R\,, \tag{4.28}$$

$$\frac{dE}{dt} = 0\,. \tag{4.29}$$

The extension consists in adding a term, $\delta RI$, to the preceding model. Repression (removal) has a deterrent effect if $\delta > 0$. In that case there is a flow out of the infective, and into the susceptible, pool. The idea is that the awareness of removals (vanishings) has the effect of driving some infectives out of the revolutionary movement. If $\delta < 0$, repression has precisely the opposite effect. Historically, both response modes are observed.[94]

[94] According to Hoover and Kowaleski, repression reduced dissent by the anti-Nazi movement in Germany (early 1940s), the human rights movement in the former USSR (1970s–1980s), and the democracy movement in Burma (late 1980s). Here, $\delta$ was positive. Repression had the reverse effect of increasing dissent by the anti-Vichy movement in France (early 1940s) and the anti-apartheid movement in South Africa (late 1980s). In these cases, $\delta$ was negative. In Hoover and Kowaleski (1992) is a review of the literature and references.

Regarding thresholds, we see from (4.27) that the revolution spreads ($\dot{I} > 0$) when the following condition is met

$$S > \frac{\alpha E_0}{r} + \frac{\delta}{r} R. \tag{4.30}$$

With $\delta = 0$, we recover the classic threshold condition (4.2), with $\alpha E_0$ as $\gamma$, just as in (4.13). If $\delta > 0$, however, more susceptibles are needed; repression deters so it is harder to initiate the revolution. If $\delta < 0$, fewer susceptibles are required; repression simply fans the flames. In such cases, a ruling elite is best advised to refrain from all action. Equivalently, revolutionaries in these situations should bend every effort to provoke brutality. The parameter, $\delta$, is what distinguishes "a revolutionary situation" from others.

We can study the dynamics of this model in $SI$ space. With a fixed total population of $N$ and fixed elite of $E_0$, let $k = N - E_0$. Then $R = k - S - I$. Substituting this into (4.26) and (4.27), we obtain

$$\frac{dS}{dt} = -rSI + (k - S - I)(\nu + \delta I), \tag{4.31}$$

$$\frac{dI}{dt} = rSI - \alpha E_0 I - \delta(k - S - I)I. \tag{4.32}$$

In the $SI$ plane, there are two equilibria[95]: $(k, 0)$, and

$$\frac{1}{\alpha\delta E_0 + \alpha E_0 + r\nu} \left(\alpha E_0(\nu + \delta k + \alpha E_0), \nu(rk - \alpha E_0)\right). \tag{4.33}$$

For illustrative parameter values, the first is a saddle, the second (4.33) is a spiral sink, and the two are connected by a heteroclinic orbit (see lecture 6).[96] This, and other trajectories, are shown in figure 4.3.

Interpreting the heteroclinic orbit politically, the equilibrium $(k, 0)$ represents a world of ideological purity—there are no revolutionaries. But, that society is "ripe for revolution;" $(k, 0)$ is a saddle, and hence unstable. The slightest subversive agitation ($I_0 > 0$) and the system will run to the spiral attractor of endemic discontent, whose exact location depends, *ceteris paribus*, on our parameter $\delta$.

[95]The second of these was obtained symbolically by *Mathematica*. See Wolfram (1991).

[96]The values employed are: $r = 0.04$, $k = 100$, $\nu = 0.45$, $\delta = 0.02$, $\alpha = 0.04$, and $E_0 = 20$.

FIGURE 4.3 Heteroclinic Social Orbit

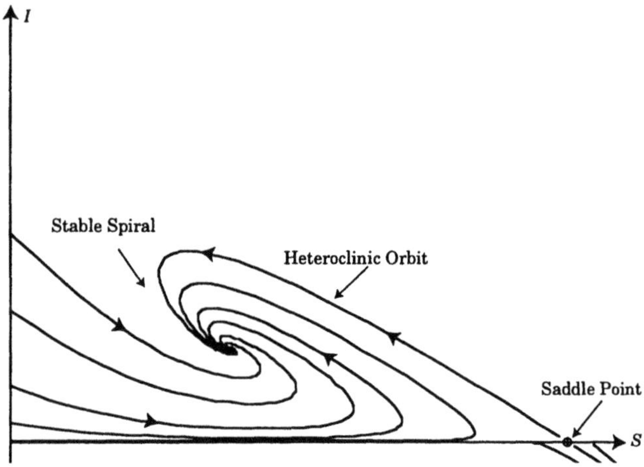


## EXTENSIONS

In point of fact, of course, the class of individuals subscribing to the elite ideology is not constant in size (at $E_0$). There is recruitment of susceptibles into the elite in addition to recruitment of susceptibles into the revolutionary class—a "struggle for the hearts and minds" of the susceptibles. And, equally important, there is often direct conflict between the power elite and the infectives; there is outright civil war. A more fully elaborated system might take the form of an ecology in which two warring (see lecture 2) and diffusing predators feed on a prey species. The resulting reaction-diffusion system would be complex mathematically, and well worth study. It would also be interesting to attempt a formulation of such a society using cellular automata (perhaps as a generalized Greenberg-Hastings model[97]) or agents.[98]

## CONCLUDING THOUGHTS

Clearly, social dynamics of fundamental interest can be generated by simple models of the sort I have advanced (very tentatively) above. Explosive upheavals; revolutions that fizzle for lack of a receptive population (e.g., $S_0 < \rho$), or that begin to spread but are reversed and crushed by an elite; longer-term cycles (undamped or

[97] See Tamayo and Hartman (1989).

[98] See Epstein and Axtell (1996).

damped) of revolutionary action; even endemic levels of social discontent, and traveling waves of revolution are all easily produced in nonlinear models. These models have the additional attractions of reflecting sensitivity to initial conditions—evident recently in Eastern Europe—and of unifying in a few variables diverse tactics of totalitarian rule and revolutionary action. Finally, even rather subtle social "bifurcation points" emerge, as where repression abruptly changes from being stabilizing ($\delta > 0$) to being inflammatory ($\delta < 0$).

Dynamical analogies are of theoretical value precisely in their power to illuminate such nonlinearities, to parsimoniously suggest general conditions under which explosive, dissipative, cyclical, even chaotic social dynamics are likely, and in so doing, to focus empirical attention on the parameters and relationships that, in fact, matter most. Indeed, without some theoretical framework, or model, it is often quite unclear what we should try to measure!

# LECTURE 5

# A Theoretical Perspective on The Spread of Drugs

## INTRODUCTION

This lecture explores another social process of considerable interest, the spread of drugs, and is divided into three parts. In Part I, a simple dynamic model of a drug epidemic in an idealized community is built up from basic assumptions concerning the interaction of subpopulations—pushers, police, and not-yet-addicted residents of the community.[99] The model combines elements of the epidemic, ecosystem, combat, and arms race models discussed above. Equilibria of the resulting dynamical system are located and classified using tools of linearized stability analysis. Trajectories are plotted for a set of initial conditions. In Part II, a spatial—reaction-diffusion—variant is presented. Then in Part III, supply, demand, and price considerations are introduced; essential, and perhaps counterintuitive, relationships

[99] Obviously, particular dynamics depend on particular drugs. No particular drug is mentioned here. We imagine an idealized drug that is totally and irreversibly addictive after some small, but hard to predict, number of uses.

between legalization, price, and crime are revealed. And in this light, the role of education is discussed.

## PART I. A DRUG EPIDEMIC MODEL

We begin with definitions and a brief discussion of variables and parameters. At any time, the population is assumed to be divided into four *disjoint* groups.

$S(t)$: The nonaddicted and susceptible population.
$I(t)$: The population of addicts, all of whom are assumed, in this simple model, to be pushers. The variable $I$ is used because this group plays a role that is mathematically analogous to the infective group in epidemiology, a parallel we shall exploit.
$L(t)$: The law enforcement, or police, force, whose sole function is assumed to be the arrest and removal of pushers.
$R(t)$: The arrested and removed, or imprisoned, population. For this simple model, removal is assumed to be permanent.

In addition to these variables, a number of parameters are involved.

$\beta$: The rate at which a contact between a pusher and a susceptible produces a new addict/pusher (price dependence is discussed in Part III below).
$\mu$: The natural growth rate in the susceptible pool, as youths come of age, say.
$\gamma$: The rate at which a contact between a pusher and a cop results in removal of the former.
$\alpha$: The rate at which an increase in pushers increases the growth rate in police. This variable reflects social alarm.[100]
$b$: The economic damping to which the police growth rate is subject.

All parameters are nonnegative real numbers.

Let us see if we cannot arrive at a plausible model by reasoning from first principles, noting connections to related phenomena as we go. Pedagogically, the exercise may illuminate the type of reasoning that often goes into the construction of models in mathematical biology, a field which, ultimately, subsumes the social sciences.

[100]Tragically, problems get more attention when they impinge on the elite than when they are confined to the ghetto. In a more realistic model, therefore, $\alpha$ would depend on the socio-economic classes into which drug abuse, and/or the crime associated with it, had spread. Here $\alpha$ is a constant.

As the simplest conceivable model, then, let us imagine that there is no population growth and no police force. At every time $t$, the population is constant at $N$ and is the sum of susceptibles $S(t)$ and pushers $I(t)$. That is,

$$N = S(t) + I(t)\,. \tag{5.1}$$

How do $S$ and $I$ evolve? Well, for a susceptible to become an addict/pusher, he or she must first come into contact with a pusher. Recognizing that real societies are heterogeneous and patchy, let us nonetheless follow the practice of theoretical epidemiology and ecology and, as a first cut, assume homogeneous mixing of pushers and susceptibles. The number of contacts is then taken to be $SI$. Of course, only some fraction $\beta$ of contacts produces new addicts. One may think of $\beta$ as the "just say no" parameter. If $\beta = 0$, every susceptible says no and there is no growth in the addicted, or "infected," pool. If $\beta = 1$, then every contact produces a new addict/pusher. On these very primitive assumptions, then, the flow out of the susceptible pool and into the addicted pool is fully described by the equations

$$\frac{dS}{dt} = -\beta SI\,, \tag{5.2}$$

$$\frac{dI}{dt} = \beta SI\,. \tag{5.3}$$

This system is none other than the most basic epidemic model, termed an "$SI$" model since the flow is strictly from susceptible to infective. Now, by virtue of (5.1), we may write $S = N - I$, and (5.3) becomes

$$\frac{dI}{dt} = \beta I(N - I) \tag{5.4}$$

whose solution is the well-known equation of logistic growth. The addicted population increases until it equals the entire population; the "epidemic" whips through the whole of society.

In fact, there are some brakes on this process. Hewing to our assumption that the drug is illegal, there is some rate at which pusher/addicts are removed from general circulation. As a first refinement on our model, let us imagine a fixed police force of size $L_0$. As in the pusher-susceptible sphere, "law of mass action" dynamics are assumed. There is homogeneous mixing of pushers and police, so that contacts proceed as $L_0 I$. And, per contact, the removal rate is $\gamma$. The idea, then, is that, as before, susceptibles flow into the addicted/pushing pool at rate $\beta SI$. But, pushers flow out of circulation and into the "removed" class at rate $\gamma L_0 I$. Since $\gamma L_0$ is just a constant, call it $\sigma$. Then we have the model:

$$\frac{dS}{dt} = -\beta SI\,, \tag{5.5}$$

$$\frac{dI}{dt} = \beta SI - \sigma I\,, \tag{5.6}$$

$$\frac{dR}{dt} = \sigma I\,. \tag{5.7}$$

Students of mathematical epidemiology will recognize this as the classic Kermack-McKendrick $SIR$ epidemic model. It is a threshold model in that susceptibles must exceed some minimum level in order for the infected, or addicted, class to grow. This is straightforward. To say the addicted class grows is to say that

$$\frac{dI}{dt} > 0,$$

which is to say that $\beta SI - \sigma I > 0$, or that

$$S > \frac{\sigma}{\beta}. \tag{5.8}$$

The ratio $\sigma/\beta$ is often termed the relative removal rate of the infection. It is the epidemic threshold. While the infection ultimately dies out—since everything eventually flows into the removed compartment—it decreases monotonically only if $S < \sigma/\beta$. Otherwise, it enjoys a period of growth—the epidemic phase—before dying out, as shown in figure 5.1, in which $\rho = \sigma/\beta$.

FIGURE 5.1 An *SIR* Epidemic Model

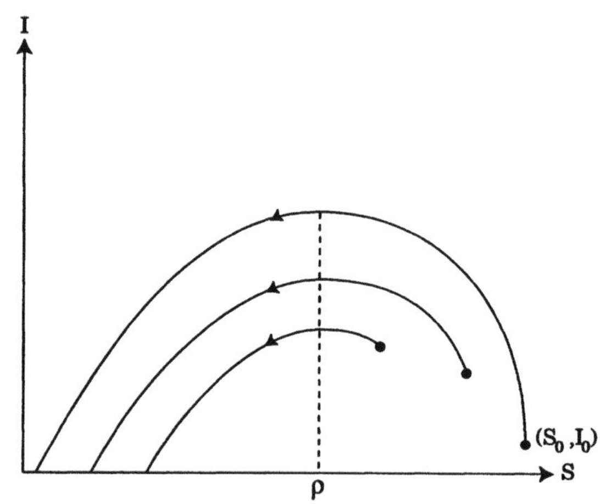


Now, from (5.5), (5.6), and (5.7), it is evident that population is still constant, since

$$\frac{dS}{dt} + \frac{dI}{dt} + \frac{dR}{dt} = 0.$$

Of course, population is not generally constant. There are so-called vital dynamics, birth and death.

As the next obvious refinement on our elementary model, then, let us assume net "births" or entrants of $\mu S$ into the susceptible cohort, where $\mu$ is the per capita growth rate. Needless to say, logistic rather than Malthusian growth is another possibility. But, keeping matters as simple as possible, we then obtain the model:

$$\frac{dS}{dt} = -\beta SI + \mu S\,, \tag{5.9}$$

$$\frac{dI}{dt} = \beta SI - \sigma I\,, \tag{5.10}$$

$$\frac{dR}{dt} = \sigma I\,. \tag{5.11}$$

Notice that (5.9) and (5.10) are the classic Lotka-Volterra predator-prey model, with pushers as predators and not-yet-addicted susceptibles as prey. Predators would die out (at rate $-\sigma I$) were there no prey to feed on (at rate $\beta SI$); and prey would flourish (at rate $\mu S$) were they not consumed (at rate $\beta SI$) by predators. Aside from the origin, this system has as its equilibrium the point $(\bar{S}, \bar{I}) = (\sigma/\beta, \mu/\beta)$, which is a center.[101] As shown in figure 5.2, the populations oscillate; the orbits are closed curves in the $SI$ phase plane.

FIGURE 5.2 Lotka-Volterra Predator-Prey Model

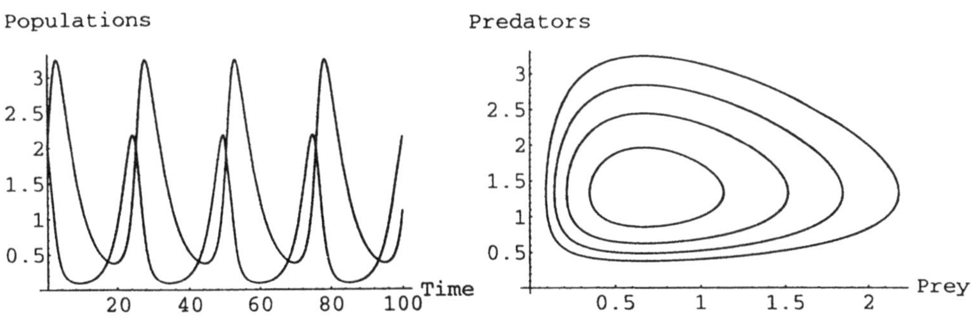


Above, we saw that in the $SIR$ epidemic model, the infected or addicted class ultimately goes to zero, as all addicts are removed. Here, we have Lotka-Volterra dynamics in which predators and prey cycle around an equilibrium. Could it be that the complete model will combine these—damped and oscillatory—tendencies in some way? We will return to this question.

[101] As developed in lecture 6, at the equilibrium, the eigenvalues of the Jacobian of (5.9)–(5.10) are imaginary. The equilibrium is nonhyperbolic and linearized stability analysis does *not* apply. But because the eigenvalues are imaginary, the equilibrium is a center or a focus; and because the system admits a Hamiltonian formulation (see lecture 4), the equilibrum is a center or a saddle. Hence, it is a center.

## THE ARMS RACE COMPONENT

Above, we defined $\sigma$ as the product $\gamma L_0$ where $L_0$ was some *fixed* level of law enforcement or police. But, the police force is not necessarily constant. So, let us relax this assumption. What, to a first order, would determine the size of the police force? Well, if no one cares about the level of addiction in society, $I(t)$, there will not be any growth. Thinking of the parameter $\alpha$ as a coefficient of societal alarm, we might posit that, without any economic damping, the police force should grow as $\alpha I$. But, as in arms race modeling, it is reasonable to assume some economic fatigue or damping, under which *rates* of growth decline the larger is the military establishment. If the damping coefficient is $b$, then the police growth rate is given by $\dot{L} = \alpha I - bL$, just as in the Richardson arms race model of lecture 3, and the complete model is as follows:

$$\frac{dS}{dt} = -\beta SI + \mu S\,, \tag{5.12}$$

$$\frac{dI}{dt} = \beta SI - \gamma IL\,, \tag{5.13}$$

$$\frac{dR}{dt} = -\gamma IL\,, \tag{5.14}$$

$$\frac{dL}{dt} = \alpha I - bL\,. \tag{5.15}$$

Notice that the term $-\gamma IL$ in (5.13) is a pusher attrition rate reminiscent of a Lanchester combat model presented in lecture 2, so this dynamical system combines elements of the epidemic, ecosystem, arms race, and combat models developed in preceding lectures. Before engaging in a linearized stability analysis of this dynamical system, let us briefly trace through the effect if, from some time, everyone says no; that is, if $\beta = 0$. Clearly, since $\beta = 0$, the growth rate in the addicted/pusher pool in (5.13) is strictly negative; this entire group is eventually removed. That being the case, (5.15) reduces to

$$\frac{dL}{dt} = -bL\,,$$

and the police force, too, "withers away," a reasonable qualitative result since the apprehension of pushers is their sole function in this model. In the end, we have a policeless society of nonaddicts and a removed population of former pushers. This little thought experiment completed, let us bring to bear some more powerful tools.

## LINEARIZED STABILITY ANALYSIS

Assuming all parameters to be positive in (5.12)–(5.15), what are the nontrivial equilibria of the system, the nonzero population levels where all derivatives are zero? We really care only about $S$, $I$, and $L$, and a bit of algebra quickly leads to the unique positive equilibrium:

$$(\bar{S}, \bar{I}, \bar{L}) = \left( \frac{\gamma}{\beta}\bar{L}, \frac{\mu}{\beta}, \frac{\alpha}{b}\bar{I} \right) . \tag{5.16}$$

Evaluated at this equilibrium (call it $\bar{x}$), the Jacobian matrix of (5.12)–(5.14), which ecologists term the community matrix, is given by

$$J(\bar{x}) = \begin{pmatrix} 0 & -\gamma\bar{L} & 0 \\ \mu & 0 & -\frac{\gamma\mu}{\beta} \\ 0 & \alpha & -b \end{pmatrix} . \tag{5.17}$$

The eigenvalues are solutions to the third-order characteristic equation

$$\mathrm{Det}(J(\bar{x}) - \lambda id) = 0 , \tag{5.18}$$

where $id$ is the identity matrix. Expanding, this characteristic equation is

$$\lambda^3 + b\lambda^2 + \left( \frac{\alpha\gamma\mu}{\beta} + \mu\gamma\bar{L} \right) \lambda + b\mu\gamma\bar{L} = 0 . \tag{5.19}$$

Equilibrium is stable if and only if the roots $\lambda_i$ of this equation have $\mathrm{Re}(\lambda_i) < 0$. For a third-degree equation,

$$\lambda^3 + a_1\lambda^2 + a_2\lambda + a_3 = 0 ,$$

the Routh-Hurwitz necessary and sufficient conditions[102] for $\mathrm{Re}(\lambda) < 0$ are

$$a_1 > 0, a_3 > 0, \text{ and } a_1 a_2 - a_3 > 0 . \tag{5.20}$$

The first two of these are obviously satisfied by (5.19), and so is the third, since

$$a_1 a_2 - a_3 = \frac{b\alpha\gamma\mu}{\beta} > 0 .$$

Therefore, the positive equilibrium in (5.16) is *stable*. There is an *endemic* level of addiction.

Earlier, I raised the question whether our simple model might somehow manifest both the damped behavior of the Kermack-McKendrick $SIR$ epidemic model and

[102] See, for example, Murray (1989, pp. 702–04), or May (1974, p. 196).

the cyclical behavior of the Lotka-Volterra predator-prey model, each of which is a special case of (5.12)–(5.15). A canonical behavior combining these would be a spiral approach to our positive equilibrium. And this is precisely the behavior we have, as shown in figure 5.3, which offers a small gallery of phase portraits. Here, the equilibrium happens to be $(\bar{S}, \bar{I}, \bar{L}) = (108, 2, 12)$.[103]

FIGURE 5.3 Drug Model Orbits and Solutions.

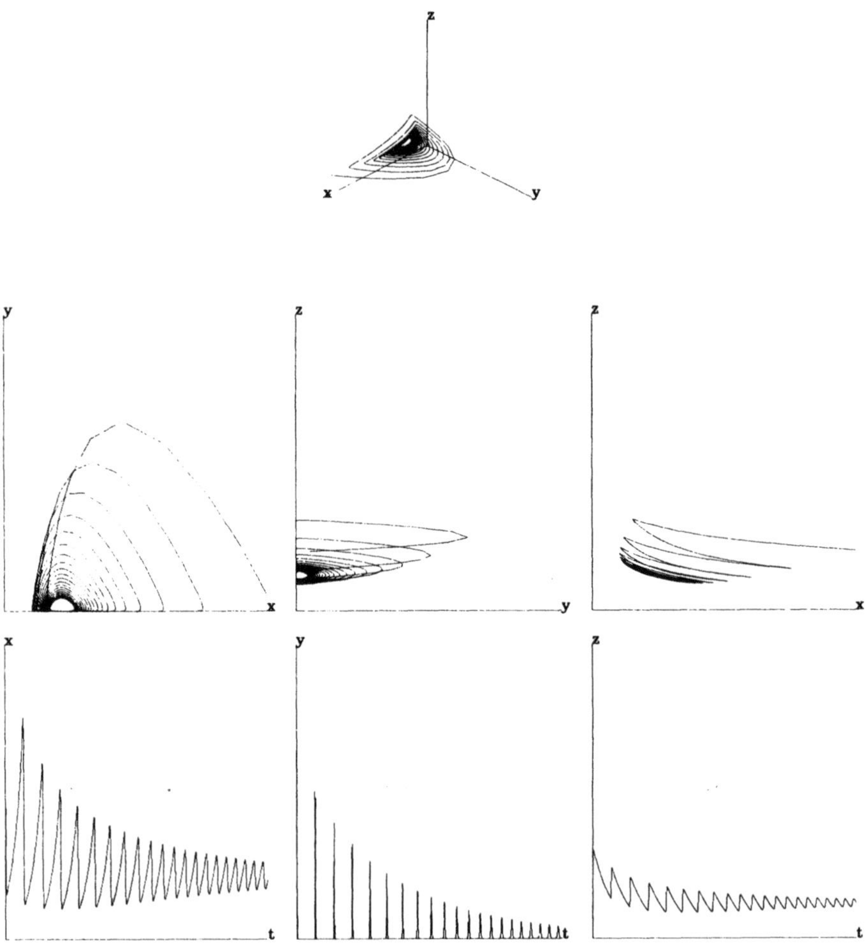


Note: Here $x$ is our $S$, $y$ is our $I$, and $z$ is our $L$.

[103]The parameter values employed are: $\beta = 0.1$, $\mu = 0.2$, $\gamma = 0.9$, $\alpha = 0.6$, and $b = 0.1$.

A natural extension is to add space. As demonstrated earlier, this can be accomplished by appending various diffusion terms to an underlying dynamic model, yielding a so-called reaction-diffusion system. One such generalization is offered next.

## PART II. DRUG WAR ON MAIN STREET: A NONLINEAR REACTION-DIFFUSION MODEL

The population is comprised of three subgroups, whose numbers and spatial distributions evolve over time. We imagine that events unfold on a one-dimensional interval—a "street." Let us define $S(x,t)$, $I(x,t)$, and $L(x,t)$ as the susceptible, infective, and law enforcement levels at street position $x$ at time $t$.[104] Denoting these functions (of $x$ and $t$) simply as $S$, $I$, and $L$, the generalized equations are as follows:

$$\begin{aligned}
\frac{\partial S}{\partial t} &= -\beta SI + \mu S + \delta_{SS}\frac{\partial^2 S}{\partial x^2}, \\
\frac{\partial I}{\partial t} &= \beta SI - \gamma IL - \delta_{SI}\frac{\partial^2 S}{\partial x^2} + \delta_{LI}\frac{\partial^2 L}{\partial x^2} + \delta_{II}\frac{\partial^2 I}{\partial x^2}, \\
\frac{\partial L}{\partial t} &= \xi SIL - bL - \delta_{IL}\frac{\partial^2 I}{\partial x^2} + \delta_{LL}\frac{\partial^2 L}{\partial x^2}.
\end{aligned} \tag{5.21}$$

Ignoring all diffusion and cross-diffusion terms, the first two equations are exactly as before. The third equation has been refined slightly. The Richardsonian damping term ($-bL$) is retained, but the first expression is now $\xi SIL$ rather than the previous $\alpha I$. The idea, recall, is that the police force grows with the level of societal alarm at the drug problem itself. This level of alarm is assumed to be a function of arrests of which the (tax-paying and police-buying) susceptibles are aware. Under our normal assumption, the arrests are proportional to $IL$, and susceptible awareness grows with exposure to these arrests, hence further multiplication by $S$, yielding the overall term $\xi SIL$.[105] These, then, are the reaction kinetics in the reaction-diffusion system (5.21).

Turning to the diffusive processes, the simplest is the susceptible case. Here, the term $\delta_{SS}(\partial^2 S/\partial x^2)$ is added, as in the models of the previous chapter, indicating that the susceptibles—while interacting with other groups—diffuse. Analogous diffusion terms appear in the equations for infectives ($\delta_{II}(\partial^2 I/\partial x^2)$) and police ($\delta_{LL}(\partial^2 L/\partial x^2)$). All these diffusivities ($\delta_{ii}$) are positive. However, the infective and police equations are more complex than this. In the police equation, there is also

[104] Here, we will not track the removed (i.e., arrested) group explicitly.

[105] We assume again that the exposuress occur through homogeneous mixing, or mass action kinetics.

a cross-diffusion term $(-\delta_{IL}(\partial^2 I/\partial x^2))$, indicating that police diffuse toward infective concentrations; they engage in "crimo-taxis," if you will. In turn, infectives (i.e., pushers) cross-diffuse in the direction of susceptibles $(-\delta_{SI}(\partial^2 S/\partial x^2))$, and cross-diffuse away from police $(\delta_{LI}(\partial^2 L/\partial x^2))$. I further assume that $\delta_{LI} > \delta_{SI}$: a pusher would rather avoid arrest than convert a susceptible to a new drug user. With all constants set,[106] the assignment of initial spatial distributions for the subpopulations is all that remains to specify the model.

Imagine, then, that everything transpires on a street 12 blocks long. At time zero, the susceptibles occupy the middle four blocks, and are 1000 strong at every point. Up at blocks 8–12 are the infectives, initially numbering but 100 at each point. And way down at blocks 1–3 are the cops, initially at token levels of 25 per point. We track the spatial evolution of each group over 50 time intervals in figure 5.4.

FIGURE 5.4 Drug War Reaction-Diffusion Model

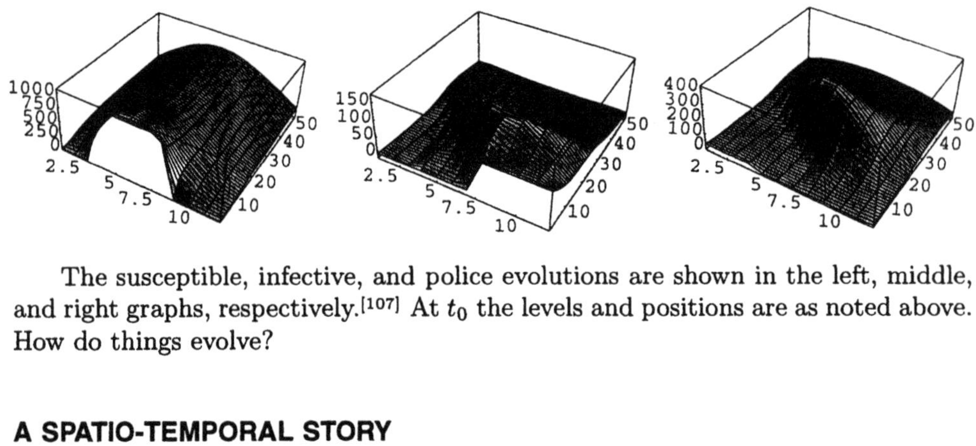


The susceptible, infective, and police evolutions are shown in the left, middle, and right graphs, respectively.[107] At $t_0$ the levels and positions are as noted above. How do things evolve?

## A SPATIO-TEMPORAL STORY

Seeing that there is a large concentration of susceptibles and few cops down the street from them, the infectives cross-diffuse to the center. Many susceptibles are converted into infectives, so the susceptible population falls and the infective one rises, now swelling with "converts" into the middle blocks. This bulging problem,

[106] Here, the values are: $\beta = 0.005$, $\mu = 0.5$, $\gamma = 0.03$, $\xi = 0.0001$, $b = 1.0$, $\delta_{SS} = 0.03$, $\delta_{II} = 0.01$, $\delta_{LL} = 0.02$, $\delta_{LI} = 0.006$, $\delta_{SI} = 0.001$, $\delta_{IL} = 0.006$.

[107] These were generated in *Mathematica* (Wolfram, 1991) using the Numerical Method of Lines. I thank Robert Axtell for his assistance.

FIGURE 5.5 Drug War Reaction-Diffusion Model: Overhead View

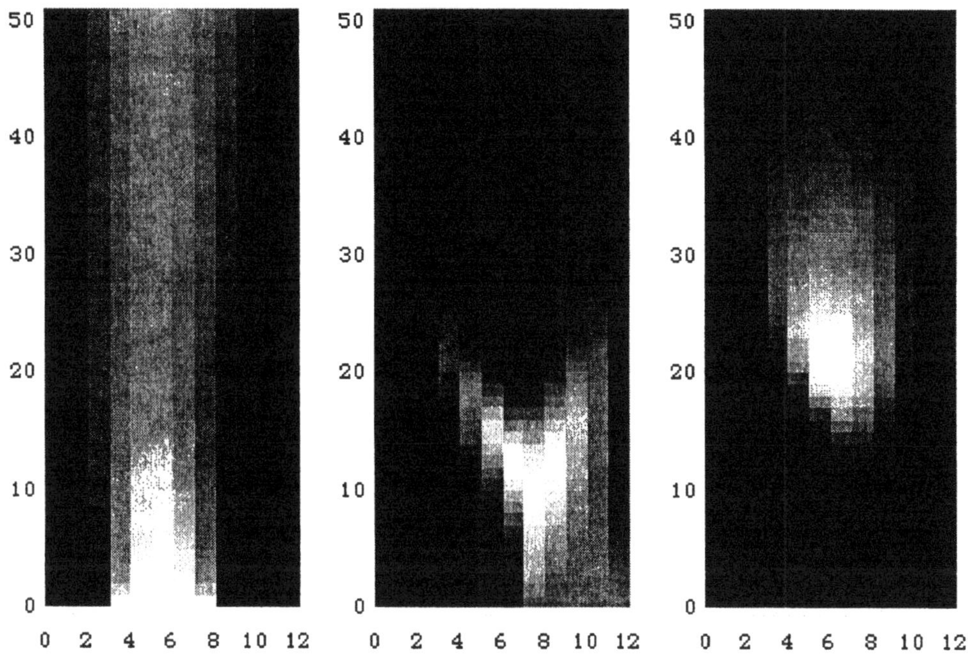


however, inspires a reaction, in the form of dramatic increases in police, who cross-diffuse from their initial barracks at the end of the street into the heart of the problem in the center. This surge in police—evident in the peak of the rightmost graph—literally scoops away the infective mound. By $t = 40$, there is hardly a problem. Hence, as before, the police "wither away" after that point, leaving the susceptibles to continue in their untroubled diffusion, as shown.

An overhead view of the same process is offered in figure 5.5. Here, the higher the numbers at a point, the lighter the shade. We can clearly see the surge of pushers, followed by the police response, the hollowing-out of the pusher mound in the center, and the withering away of the police.

A nonlinear reaction-diffusion model allows us to generate a plausible spatio-temporal story of basic interest.[108]

[108]My point here is that there exist parameter values and initial conditions under which the spatio-temporal story emerges. A seperate study would examine the robustness of this result under a wide range of parameter and initial values.

## PART III. ELASTICITIES AND THE TWO COMPONENTS OF DEMAND

Economists will have recognized one glaring oversight in the models thus far developed—the price of the drug is not represented. Clearly, $\beta$—the rate at which a contact between a pusher and a *not-yet-addicted* susceptible results in a use—depends on price. *Ceteris paribus*, the likelihood that someone who is not yet addicted will "just say no" should rise ($\beta$ should fall) with price. So, for that group we assume

$$\beta'(P) < 0. \tag{5.22}$$

If we further assume price to be determined by supply and demand, then, in principle, the effect of drug interdiction operations on first uses of the drug can be gauged, as shown in figure 5.6. Interdiction shifts supply in (further discussion below), increases the equilibrium price, reduces $\beta$, and in principle changes the equilibrium level of drug consumption (from $Q^0$ to $Q^i$).

By contrast, within the addicted population, the quantity demanded is not sensitive to price; the demand curve is a vertical line, as shown in figure 5.7.

FIGURE 5.6 Interdiction Increasing Price

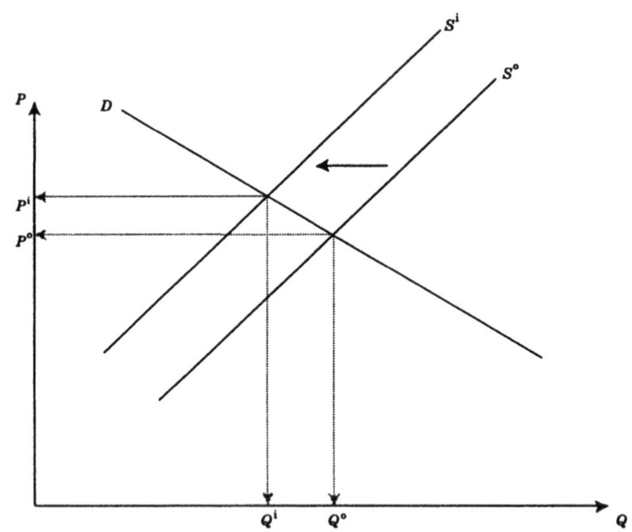

FIGURE 5.7 Addicted Demand

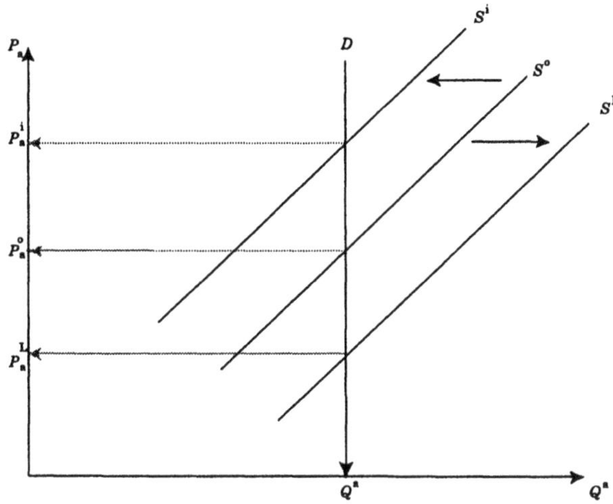

In such a market, supply interdiction merely increases price with no effect on the equilibrium drug quantity. The result is simply more street crime, as addicts do whatever is necessary (e.g., rob and loot) to come up with the price.

Now, figure 5.6 assumes the market to be comprised only of not-yet-addicted susceptibles, while figure 5.7 assumes it to be comprised only of addicts. In fact, both groups are present. These two components of total demand are represented in figure 5.8.

FIGURE 5.8 A Drug Market

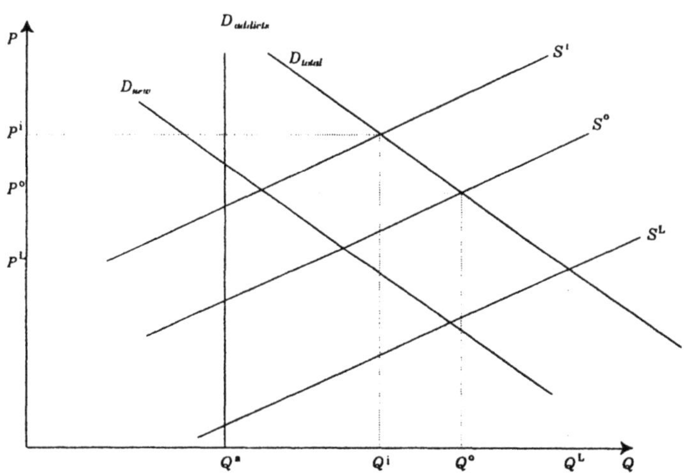

The diagram encompasses an addicted pool, with its distinguishing feature, a vertical demand curve ($D_{addicts}$), representing complete inelasticity of demand with respect to price. The addicted group has "got to have" $Q^a$, and that's that. The $D_{new}$ curve represents the nonaddicted population, with its downward-sloping demand curve. Here, price does affect the demand by potential first users of our drug. Total demand ($D_{total}$) is the horizontal sum of these two components. Finally, we imagine $S^0$ to represent the current supply curve in this economy. Then the equilibrium price and quantity in the market is $(P^0, Q^0)$.

Let us use this framework to compare the short-term effect of two supply-oriented policies, returning later to the longer-term issue of demand adjustment. The two policies are, of course, interdiction/prohibition and legalization.[109] Assuming (perhaps generously) that interdiction is technically feasible, it shifts supply inward to $S^i$. Legalization—whose aim is to drive, underbid, the cartels out of business—would shift supply outward to $S^L$.

## INTERDICTION

First, under interdiction the equilibrium price rises to $P^i$. Since addicts have "got to have it," they will do whatever is necessary to raise the money required to support their habits, and we can expect an increase in street crime. In the nonaddicted group, the quantity demanded falls as price increases. So, interdiction produces *fewer first users but more crime.*

## LEGALIZATION

Legalization shifts supply out to $S^L$. The equilibrium price falls to $P^L$, so addicts need less money than they did before (at $S^0$) to satisfy their habits; not all of the crime in which they had been engaged would now be necessary, and we would expect to see street crime decrease. New users, however, may be expected to increase (and to increase their consumption) as experimentation becomes cheaper (and, obviously less risky legally) under legalization. So, this policy produces *less crime, but more new users.* All of this is encapsulated in the purely heuristic diagram in figure 5.9.

The debate is, predictably, wide open because no one really knows much about this curve, about the factors underlying it—price elasticities, propensities to engage in crime, effectiveness of interdiction, and so on—or about the social welfare to be associated with different points on the curve. As a result, the most fundamental issue is open: *Do we want drug prices to rise dramatically or fall dramatically?*

But, at least the analysis reveals those factors about which advocates of the different policies are, in fact, disagreeing.[110] A modest claim, to be sure.

[109] For a thorough historical discussion and policy analysis, see Stares (1996).

[110] "Just saying no" reduces crime indirectly because $D_{total}$ shifts left, reducing the equilibrium price, and hence the incentive to engage in street crime.

FIGURE 5.9 The Legalization-Interdiction Trade-off

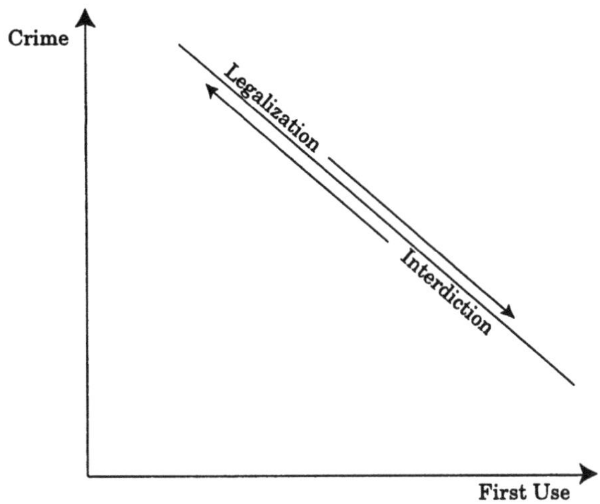


## EDUCATION AND $\beta$ FOR THE NONADDICTED POPULATION

A further point concerns demand among the nonaddicted susceptibles, where the likelihood that someone will "just say no" probably depends on prices. While the incorporation of supply, demand, and price into the model allows us to connect, through $\beta$, the "epidemic dynamics" developed earlier to the alternative policies, making $\beta$ a function *solely* of price is not completely satisfying, for the following reason: Imagine an all out, two-front "war on drugs" that simultaneously cut supply through interdiction and cut demand (here in the new-use sense) through education. The equilibrium price could be left unaffected. If $\beta$ were a function solely of price, then it, too, would be unaffected, and the drug epidemic dynamics would be exactly as they were before, which seems implausible. One would therefore want $\beta$ to depend on price *and* education ($E$) in some (perhaps parametric) way such that

$$\frac{\partial \beta}{\partial P} < 0 \text{ and } \frac{\partial \beta}{\partial E} < 0 . \tag{5.23}$$

Many "toy" candidates for $\beta(P, E)$, could be constructed for simulation purposes. But, basic empirical work is essential here, as elsewhere. Indeed, it would seem that education is the prime way to avoid a very disturbing long-run scenario under legalization.

## LEGALIZATION, EDUCATION, AND THE LONG RUN

Specifically, a potentially serious problem is that with repeated use over time, the price elasticity of demand for individuals initially in the nonaddicted pool will go toward zero: they may become addicts.

FIGURE 5.10 Legalization and Long-Run Price

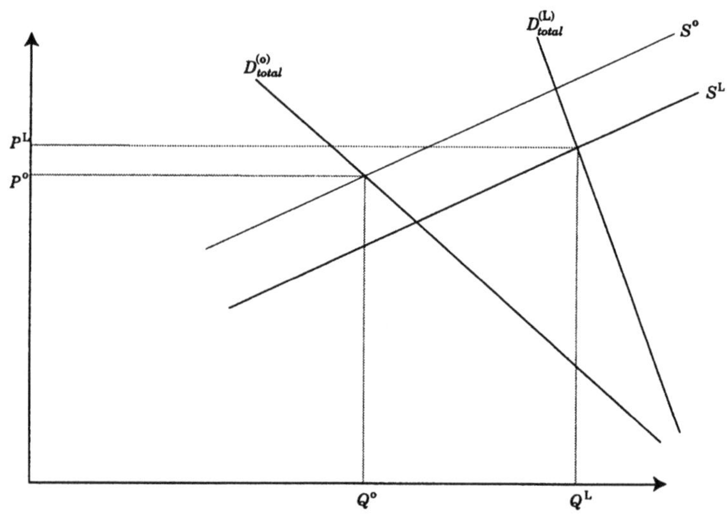


In turn, we can expect increasing levels of addiction (a) to increase the total quantity demanded at any price and (b) to reduce the price elasticity of aggregate demand. The total demand curve, thus, would be expected to shift rightward *and become more vertical*, as shown in figure 5.10. Counterintuitively, in this case the equilibrium price could rise despite increased supply. That is, we could have ($P^L > P^0$), as shown. Street crime, driven by rising drug prices, could in principle actually increase under legalization! The magnitude and direction of price changes, it must be emphasized, would depend on the supply-response to legalization, the rate at which new users become addicts, and so forth.

But, the obvious way to undercut this type of evolution—whose likelihood I do not claim to estimate—is education. The aim of education, in a technical economic sense, would be to "flatten" the individual's indifference curves (between drug consumption and engagement in other forms of recreation) such that there is no point of tangency between the indifference curve and the budget constraint, leaving only the corner solution, "just say no," as feasible.[111] The production of horizontal indifference curves for fur coats is the aim of animal rights activists, for example; there is then no pair of positive prices—for real and imitation fur—at which the indifference curve is tangent to the individual's budget constraint, and he or she "just says no" to fur.[112]

In any event, it seems clear that, if legalization is to avoid the long-run problem suggested above, it may be necessary to increase education very substantially.

## CLOSING THOUGHTS

It should be emphasized that, in addition to all sorts of implicit assumptions concerning the factors discussed above, positions on legalization reflect fundamental attitudes on the appropriate role of government in regulating individual choices generally, a crucial question not addressed here.

Obviously, this theoretical exercise does not purport to *resolve* any policy issue. Rather, it seeks to focus the debate by helping to identify the empirical issues that deserve highest priority, by encouraging explicitness in the statement of assumptions, and by offering very preliminary, but testable, models of the dynamic interaction of core variables.

Finally, from a purely intellectual standpoint, the discussion suggests the relevance to social science of seemingly distant areas like mathematical epidemiology and ecosystem modeling. And it tries to illustrate how simple nonlinear models are built and analyzed.

[111]The idea that the individual will consume where the indifference curve and budget constraint are tangent is developed as follows. We imagine that an individual derives utility from the consumption of quantity $q_d$ of drugs and quantity $q_j$ of some alternative commodity, and that total utility is a function (with all nice behaviors) of these, $u(q_d, q_j)$. On an indifference, or isoutility, curve, $du \equiv 0$, so that (i) $(\partial u/\partial q_d)dq_d + (\partial u/\partial q_j)dq_j = 0$, meaning that the slope of the indifference curve in $(q_d, q_j)$ space is given by (ii) $(dq_d/dq_j) = -(\partial u/\partial q_j)/(\partial u/\partial q_d)$. But if the individual maximizes utility subject to a budget constraint $B = P_d q_d + P_j q_j$, then at the constrained maximum, (ii) must equal the price ratio, making the indifference curve tangent to the budget constraint. To see this, form the Lagrangian for the constrained problem: $L = u(q_1, q_2) - \lambda(B - P_d q_d - P_j q_j)$. The first-order conditions for a maximum being $(\partial L/\partial q_d) = (\partial L/\partial q_j) = 0$, we immediately obtain that $(\partial u/\partial q_d) = \lambda P_d$ and $(\partial u/\partial q_j) = \lambda P_j$, making the right-hand side of (ii) equal the price ratio on division. Obviously, for the tangency point to be unique, the indifference curves must be strictly convex, which is among the behaviors of a nice $u$-function.

[112]Indeed, one might go farther and argue that the goal is to ensure—through education—that indifference curves for harmful commodities are positively sloped. In other words, people would have to be compensated with higher quantities of some other good to induce them to consume the harmful commodity. I thank Steven McCarroll for this idea.

# LECTURE 6

# An Introduction to Nonlinear Dynamical Systems

In this lecture, I want to collect some basic, and very powerful, results from the qualitative theory of nonlinear autonomous differential dynamical systems, primarily in the plane. In a field as vast as nonlinear dynamics, any essay of the present length must be selective. In this case, the story begins with linearized stability analysis for hyperbolic equilibria and proceeds to develop some diagnostic tools for nonhyperbolic cases (including the use of polar coordinates, Lyapunov functions, and Hamiltonian formulations). The distinctly nonlinear phenomenon of the limit cycle is then discussed and Hilbert's—still unsolved—16th Problem is stated. The Poincaré-Bendixson and Hopf Bifurcation Theorems are presented, as well as an introduction to Poincaré maps, which beautifully connect the world of continuous systems (flows) to that of discrete systems (maps). Tools for precluding periodic orbits—the Bendixson and Bendixson-Dulac negative tests—are then presented and applied to a Kolmogorov system which, naturally, provides a forum for Kolmogorov's Theorem on cycles. Powerful as they are, none of these methods give much insight concerning how the local equilibria and limit cycles fit together globally—in the phase plane as a whole. Index theory penetrates deeply into this question, to reveal topological "conservation laws" of great interest. I present some of the fundamental results in this area, and give an index theoretic proof of Brouwer's famous fixed point theorem on the disk. Extending these ideas to closed surfaces (two-manifolds) like the sphere and torus, the lecture concludes with an informal presentation of the celebrated Poincaré-Hopf Index Theorem from differential topology.

## NONLINEAR AUTONOMOUS PLANAR SYSTEMS

We consider the nonlinear autonomous system[113]

$$\begin{aligned} \dot{x}_1 &= f_1(x_1, x_2), \\ \dot{x}_2 &= f_2(x_1, x_2), \end{aligned} \tag{6.1}$$

where $f_1$ and $f_2$ are $C^1$ on $\mathcal{R}^2$. The system (6.1) defines a vector field $F = (f_1(x_1, x_2), f_2(x_1, x_2))$ from $\mathcal{R}^2$ to $\mathcal{R}^2$. Recall that the Jacobian of $F$ at a point $x$ is the matrix

$$DF(x) = \begin{pmatrix} \frac{\partial f_1}{\partial x_1}(x) & \frac{\partial f_1}{\partial x_2}(x) \\ \frac{\partial f_2}{\partial x_1}(x) & \frac{\partial f_2}{\partial x_2}(x) \end{pmatrix}. \tag{6.2}$$

In the linear homogeneous case where $F(x) = Ax, DF(x)$ is just $A$. One major difference between linear and nonlinear systems in the plane is that the latter may

[113]Much of the orbit theory developed below rests on the assumption that the nonlinear first-order initial value problem—$dy/dx = f(x, y), f$ continuous on $\Omega = [a, b] \times [c, d], y(x_0) = y_0$—has a unique solution $y$, defined and continuous on a closed subinterval of $[a, b]$ containing $x_0$. A basic theorem is that if $f$ is Lipschitz on $\Omega$, such is the case. A full discussion would require development of metric space results quite foreign to the rest of this essay, notably (a) that the space $c[a, b]$ of functions continuous on an interval $[a, b]$—the space that would contain any solution—is complete with metric

$$\rho(x, y) = \max_{t \in [a,b]} |x(t) - y(t)|;$$

and (b) Banach's theorem that a contraction mapping on a complete metric space has a unique fixed point. These theorems in hand, however, existence-uniqueness is direct. First, one observes that $y$ is the unique solution to the initial value problem above if and only if it is the unique fixed point of the integral operator, $P : c[a, b] \to c[a, b]$ defined by

$$P[y] = y_0 + \int_{x_0}^{x} f(s, y(s))ds.$$

The proof then consists in establishing simply that if $f$ is Lipschitz on $\Omega$, then $P$ (for Picard) is a contraction on $c[a, b]$. As Kreyszig puts it, "the idea of the approach is quite simple: [the initial value problem] will be converted to an integral equation which defines a mapping $T$, and the conditions of the theorem will imply that $T$ is a contraction such that its fixed point becomes the solution of our problem." Our $P$ is just Kreyszig's $T$. For detailed proofs of all the many subsidiary claims involved, see, for example, Kreyszig (1978), Naylor and Sell (1982), Waltman (1986), Groetsch (1980), Gamelin and Greene (1983), or Marsden (1974).

have multiple equilibria whereas the linear homogeneous system $\dot{x} = Ax$ has the origin as its unique equilibrium.

Suppose now that $\bar{x} = (\bar{x}_1, \bar{x}_2)$ is an equilibrium of the nonlinear system (6.1). Assuming $f_1$ and $f_2$ have partial derivatives of all orders on an open set containing $\bar{x}$, we can expand each in a Taylor series about $\bar{x}$. Retaining only the linear terms, we obtain the so-called linear variational equations:

$$\begin{aligned} \dot{x}_1 &= f_1(\bar{x}_1, \bar{x}_2) + \frac{\partial f_1}{\partial x_1}(\bar{x}_1, \bar{x}_2)(x_1 - \bar{x}_1) + \frac{\partial f_1}{\partial x_2}(\bar{x}_1, \bar{x}_2)(x_2 - \bar{x}_2), \\ \dot{x}_2 &= f_2(\bar{x}_1, \bar{x}_2) + \frac{\partial f_2}{\partial x_1}(\bar{x}_1, \bar{x}_2)(x_1 - \bar{x}_1) + \frac{\partial f_2}{\partial x_2}(\bar{x}_1, \bar{x}_2)(x_2 - \bar{x}_2). \end{aligned} \tag{6.3}$$

Since $(\bar{x}_1, \bar{x}_2)$ is an equilibrium of (6.1), we have $f_1(\bar{x}_1, \bar{x}_2) = f_2(\bar{x}_1, \bar{x}_2) = 0$. Defining the deviations $y_1 = (x_1 - \bar{x}_1)$ and $y_2 = (x_2 - \bar{x}_2)$, (6.3) becomes

$$\dot{y} = DF(\bar{x})y, \tag{6.4}$$

the familiar linear homogeneous problem. If, for expository ease, we assume $DF(\bar{x})$ to have distinct real nonzero eigenvalues, $\lambda_1$ and $\lambda_2$, with (perforce linearly independent) eigenvectors $\nu_1$ and $\nu_2$, the general solution of (6.4) is

$$\begin{pmatrix} y_1 \\ y_2 \end{pmatrix} = c_1 e^{\lambda_1 t} \nu_1 + c_2 e^{\lambda_2 t} \nu_2 . \tag{6.5}$$

Given the vector initial condition $y(0) = y_0$, we can determine the $c$-values; indeed, with the matrix exponential, we obtain the succinct form:

$$y = e^{DF(\bar{x})t} y_0 . \tag{6.6}$$

You will doubtless recall from previous work that if both eigenvalues are negative reals, or have negative real part in the complex (conjugate) case, then the origin is a globally asymptotically stable equilibrium of (6.4).[114] If both eigenvalues are, in fact, negative (positive) reals, then the origin is a stable (unstable) *node*. In the complex case, if real parts are negative (positive), then the origin is a stable (unstable) *spiral*. It is a *center* if the eigenvalues are purely imaginary, and a *saddle* if they are real with $\lambda_1 < 0 < \lambda_2$. All of this is summarized in figure 6.1, whose axes are $\gamma = Tr[DF(\bar{x})]$ and $\beta = \text{Det}[DF(\bar{x})]$.

You doubtless also recall that there are the further, repeated eigenvalue, cases where issues of multiplicity arise.[115] As an exercise, you might enjoy demonstrating

[114] On stability and asymptotic stability, see Hirsch and Smale (1974; Section 9.2).

[115] See for example, Hirsch and Smale (1974), Waltman (1986), Braun (1983), or Borelli and Coleman (1987).

formally a handy fact that emerges from this figure, namely that the origin is stable if the Trace is negative and the Determinant is positive.

FIGURE 6.1 Global Stability of the Origin in the Linear Case (Distinct Nonzero Eigenvalues)

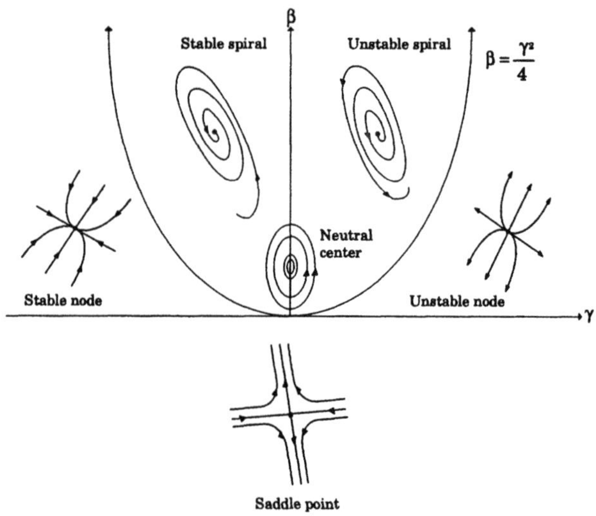


Source: Based on Edelstein-Keshet (1988, p. 190).

Of course, all of this has to do with (6.4), when we are really interested in (6.1). If, somehow, we knew that the behavior of the linearization (6.4) faithfully represented the behavior of (6.1) in the vicinity of $\bar{x}$, then we could conclude from the *global* stability of (6.4) at the origin, the *local* stability of (6.1) at $\bar{x}$. Indeed, at each equilibrium $\bar{x}$ of (6.1), we could simply inspect the eigenvalues of the Jacobian $DF(\bar{x})$ and classify exactly as in the linear case. Amazingly, for a certain class of equilibria we can do just that! With that thought in mind, we make the following definition.

**Definition.** An equilibrium $\bar{x}$ of the (linear or nonlinear) system $\dot{x} = F(x)$ is *hyperbolic* if and only if all eigenvalues of the Jacobian evaluated at $\bar{x}$, $DF(\bar{x})$, have nonzero real parts.

Now it is a very useful fact that for hyperbolic equilibria, linearization is reliable. If $\bar{x}$ is a hyperbolic equilibrium of (6.1), then its type (node, focus, saddle) and its stability correspond exactly to the type and stability of the zero equilibrium of (6.4), its local linearization. This is a consequence of the Hartman-Grobman Theorem.

**Theorem 1** (Hartman-Grobman). *If $\bar{x}$ is a hyperbolic equilibrium of $\dot{x} = F(x)$, then there is a neighborhood of $\bar{x}$ in which $F$ is topologically equivalent to the linear vector field $\dot{x} = DF(\bar{x})x$.*

Here, two dynamical systems $\dot{x} = f(x)$ and $\dot{x} = g(x)$, defined on open sets $U$ and $V$ of $\mathcal{R}^2$ are *topologically equivalent* if there exists a homeomorphism (a one-to-one continuous mapping with continuous inverse) $h : U \to V$ mapping the orbits of $f$ onto those of $g$ and preserving direction in time.

This gives us a "recipe," if you will, for local stability analysis of hyperbolic equilibria of nonlinear systems: For each such equilibrium $\bar{x}$, compute the eigenvalues of the Jacobian, $DF(\bar{x})$, and classify as you would classify the origin in the linear system $\dot{x} = DF(\bar{x})x$. Neighborhood stability analysis, as the technique is also known, is among the most ubiquitous tools in all of applied mathematics. It appears frequently in physics and engineering and is a cornerstone of mathematical ecology, epidemiology, and population genetics. Linearized stability analysis also pervades important subfields of mathematical economics, evolutionary game theory, and other areas of social science, including the theory of arms races, wars, and revolutions, which topics I discuss in other lectures. Precisely because examples are so abundant in the literature, I thought a "nonstock" application would be nice. A particularly ingenious one arises in connection with traveling wave solutions to reaction-diffusion equations. Since these figure in two other lectures of mine (on revolutions and drugs), I thought "kill two birds with one stone."

## TRAVELING WAVES, HETEROCLINIC ORBITS, AND LINEARIZED STABILITY ANALYSIS.

Among the most famous reaction-diffusion equations is Fisher's (1937) equation governing the spread of an advantageous gene in a population. With $D$ a diffusion constant, $\alpha$ a parameter measuring the intensity of selection, and $p(x,t)$ the gene's frequency at point $x$ at time $t$, Fisher's equation is

$$\frac{\partial p}{\partial t} = D\frac{\partial^2 p}{\partial x^2} + \alpha p(1-p)\,. \tag{6.7}$$

We wish to establish whether (6.7) admits traveling wave solutions consistent with biologically realistic assumptions. We posit traveling wave solutions of the form[116]

$$P(z) = p(x,t) \text{ with } z = x - ct\,.$$

[116]This discussion parallels Edelstein-Keshet (1988). See also Murray (1989).

By the chain rule,

$$\frac{\partial P}{\partial t} = \frac{\partial P}{\partial z}\frac{\partial z}{\partial t} = -c\frac{\partial P}{\partial z} \text{ and}$$
$$\frac{\partial P}{\partial x} = \frac{\partial P}{\partial z}\frac{\partial z}{\partial x} = \frac{\partial P}{\partial z}, \text{ so (6.7) becomes}$$
$$-c\frac{dP}{dz} = D\frac{d^2 P}{dz^2} + \alpha P(1-P)\,, \tag{6.8}$$

a nonlinear second-order ordinary differential equation. Defining $-S = dP/dz$, we recast (6.8) as the system

$$\begin{aligned} \frac{dP}{dz} &= -S\,, \\ \frac{dS}{dz} &= \frac{\alpha}{D}P(1-P) - \frac{c}{D}S\,. \end{aligned} \tag{6.9}$$

Notice that $z$ now plays the role normally played by $t$, so that, in phase space, $z$ is changing along—is parametrizing—trajectories and can assume all real values. By contrast, we place some definite conditions on $P$. First, since $P$ is the relative frequency of our advantageous gene, we require that it be bounded:

$$\text{(i)}\ \ 0 < P(z) < 1 \text{ for all } z\,, (-\infty < z < +\infty)\,.$$

Second, when we say the gene is advantageous, we mean that it will ultimately dominate the pool. In other words,

$$\text{(ii)}\ \ P(z) \to 1 \text{ as } z \to -\infty (\text{since } t \to \infty)\,.$$

Symmetrically, and finally, we assume that

$$\text{(iii)}\ \ P(z) \to 0 \text{ as } z \to +\infty\,.$$

Now, the dynamical system (6.9) has two equilibria in the $PS$ plane: $\bar{x}_1 = (P_1, S_1) = (0,0)$ and $\bar{x}_2 = (P_2, S_2) = (1,0)$. Thus, $P = 0$ at $\bar{x}_1$ and $P = 1$ at $\bar{x}_2$. Hence, if we are to satisfy conditions (ii) and (iii), we need an orbit *connecting* these equilibria. Such orbits are termed *heteroclinic.* Technically, a heteroclinic orbit is one whose $\alpha$ and $\omega$ limit sets (defined below) are distinct equilibrium points.

Two centers clearly cannot be heteroclinic. Neither can a pair of sinks (attractors) or a pair of sources (repellors). Rather, we need one unstable and one stable equilibrium, so that the trajectory emanating from one may be attracted to the other.

We will know that Fisher's equation (6.7) admits a traveling wave solution precisely if we can show that (6.9) admits a certain heteroclinic orbit. Enter linearized stability analysis!

The Jacobian of (6.9) is

$$DF(x) = \begin{pmatrix} 0 & -1 \\ \frac{\alpha}{D}(1-2P) & \frac{-c}{D} \end{pmatrix}.$$

At $\bar{x}_1$,

$$DF(\bar{x}_1) = \begin{pmatrix} 0 & -1 \\ \frac{\alpha}{D} & \frac{-c}{D} \end{pmatrix},$$

and at $\bar{x}_2$,

$$DF(\bar{x}_2) = \begin{pmatrix} 0 & -1 \\ \frac{-\alpha}{D} & \frac{-c}{D} \end{pmatrix}.$$

Now, the eigenvalues in each case are functions of the parameters $\alpha$, $c$, and $D$. One can easily show, by computing the eigenvalues, that if

$$c > 2\sqrt{\alpha D}, \tag{6.10}$$

the origin, $\bar{x}_1$, is a stable node ($\lambda_1, \lambda_2 < 0$) and $\bar{x}_2$ is a saddle point ($\lambda_1 < 0 < \lambda_2$), which is just the sort of configuration we need. These equilibria are indeed connected by the desired heteroclinic orbit, shown in figure 6.2.[117]

FIGURE 6.2 Heteroclinic Orbit

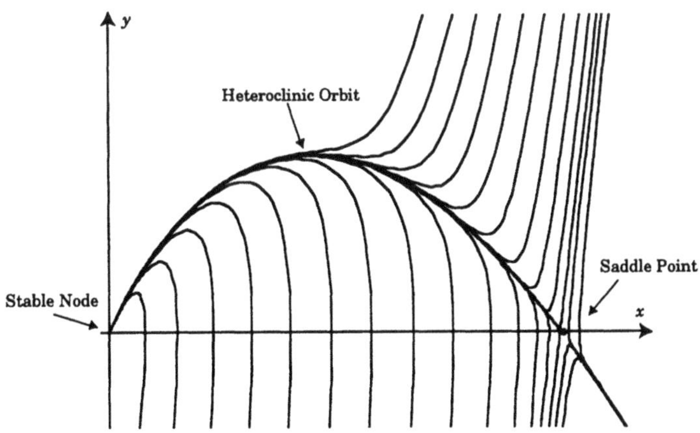


[117]To be precise, the fact that the system (6.9) has exactly one stable and one unstable equilibrium does not alone establish that the two are connected by a heteroclinic orbit. The unstable equilibrium could be surrounded by a semi-stable limit cycle (see below). In fact, for the system (6.9), this can be ruled out by Bendixson's negative test (see below).

Finally, interpreting (6.10), we have also learned that the traveling wave's speed is bounded below by $2\sqrt{\alpha D}$, a very nice byproduct of the analysis. This value is a "bifurcation" point (see discussion below) also, in that the phase portraits for $c \leq 2\sqrt{\alpha D}$ and for $c > 2\sqrt{\alpha D}$ are not topologically orbitally equivalent; for $c \leq 2\sqrt{\alpha D}$, we lose our heteroclinic orbit.

There are a number of things to relish in this example, quite aside from the mathematical propagating wave itself. The reasoning, originally due to Fisher, Kolmogorov, and others, is a marvel of indirection. We start with a nonlinear partial differential equation which we never solve; we posit a traveling wave solution whose substitution into the original equation converts it into a nonlinear system of ordinary differential equations, which we also never solve. Rather, we make a small number of reasoned assumptions about the bounds and asymptotic behavior of $P$, deduce that there must be a heteroclinic orbit, and use linearized stability analysis to establish the requisite parameter relationship, which happens to provide the minimum wave speed as a byproduct!

For a fuller account of possibilities under the more general equation

$$\frac{\partial p}{\partial t} = \frac{\partial^2 p}{\partial x^2} + f(p)\,, \tag{6.11}$$

see Fife (1979), Britton (1986), and Smoller (1983).

Finally, before leaving this example, notice how the nonlinearity of $f$ is essential to the traveling wave solution. (For instance, if you again begin by positing such a wave solution and try to carry through the same derivation as above, but now with $f$ linear, you will obtain a linear analogue of the system (6.9), with a *single* equilibrium and, hence, no prospect of a heteroclinic orbit.)

Returning to the main plot, however, the central point is that linearization is an extremely powerful tool whose applications are really quite far flung, as this example suggests. Powerful as it is, linearized stability analysis is not omnipotent. If it were, nonlinear dynamics would be a pretty small field.

## WHEN LINEARIZATION FAILS: SOME NONHYPERBOLIC EQUILIBRIA

As an example of a failure of linearization, consider the system[118]

$$\begin{aligned} \dot{x}_1 &= x_2 + a x_1 (x_1^2 + x_2^2)\,, \\ \dot{x}_2 &= -x_1 + a x_2 (x_1^2 + x_2^2)\,. \end{aligned} \tag{6.12}$$

[118] Hale and Koçak (1991, p. 334).

What is the nature of the zero equilibrium? Applying the recipe, we compute the Jacobian matrix and evaluate at $\bar{x} = 0$ to yield:

$$DF(\bar{x}) = \begin{pmatrix} 0 & 1 \\ -1 & 0 \end{pmatrix} . \tag{6.13}$$

The linearized system at $\bar{x}$ is thus the classic harmonic oscillator familiar from elementary physics: $\dot{x}_1 = x_2; \dot{x}_2 = -x_1$. The characteristic polynomial is

$$\lambda^2 + 1 = 0 \tag{6.14}$$

whose solutions are the purely imaginary eigenvalues $\lambda_{1,2} = \pm i$.

Since the eigenvalues have zero real parts, they are *non*hyperbolic. Now, in the linear case, imaginary eigenvalues would indicate a center, neutral stability (like the harmonic oscillator). Is that true here?

If we convert to polar coordinates, the system (6.12) becomes

$$\begin{aligned} \dot{\theta} &= -1 , \\ \dot{r} &= ar^3 , \end{aligned} \tag{6.15}$$

with equilibrium $\bar{r} = 0$. Clearly $\bar{r}$ is globally asymptotically stable if $a < 0$, neutrally stable (a center) if $a = 0$, and unstable if $a > 0$, as indicated in figure 6.3.

FIGURE 6.3 Phase Portraits for (6.12)

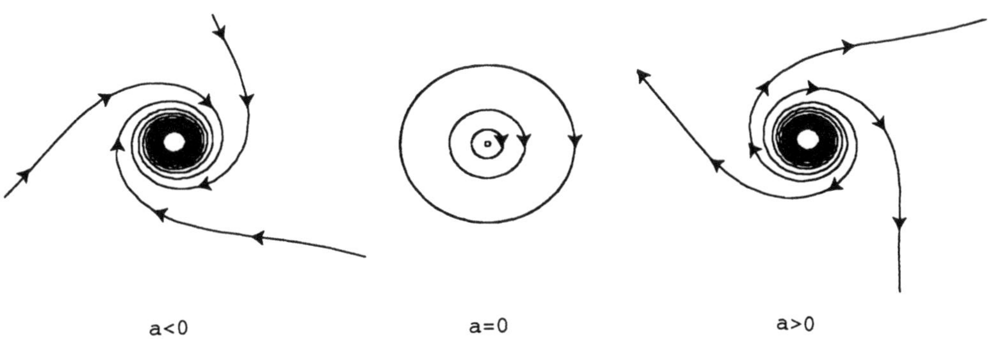


Clearly, then, as a general proposition, linearization is *not* reliable for *non*hyperbolic fixed points. However, for critical points with purely imaginary (as against, for example, zero) eigenvalues, we do have the following result of Poincaré's.

**Theorem 2** (Poincaré). *A center equilibrium of the linearized system (6.4) is either a center or a focus of the original nonlinear system (6.1).*[119]

This is understandable in light of our preceding results. Since the eigenvalues are imaginary, nodes and saddles are precluded. We have just displayed a focus. So a constructive proof will be done once we find a center equilibrium of a nonlinear system that is also a center of its linearization (try the Lotka-Volterra predator-prey model, equations (4.6) above).

As a methodological point, this example illustrates the usefulness of polar coordinates in some cases (the presence of the term $x_1^2 + x_2^2$ is always suggestive). Another way to analyze the nonhyperbolic equilibrium $\bar{x} = 0$ of (6.12) is to reason as follows. We are worrying about whether the representative point on the solution $(x_1(t), x_2(t))$ moves *toward* the equilibrium, the origin, over time. So let us look at the Euclidean distance

$$||x|| = \sqrt{x_1^2 + x_2^2}$$

or, equivalently, the square of the distance, since if the square approaches zero, so must the distance itself. Accordingly, define

$$V = ||x||^2 = x_1^2 + x_2^2 \,. \tag{6.16}$$

How does $V$ change with time? By the chain rule:

$$\dot{V} = \frac{\partial V}{\partial x_1}\dot{x}_1 + \frac{\partial V}{\partial x_2}\dot{x}_2 \,.$$

But, with $\dot{x}_1$ and $\dot{x}_2$ from (6.12), we have

$$\dot{V} = 2a(x_1^2 + x_2^2)^2 \,,$$

which agrees with our previous result: the equilibrium ($\bar{r} = 0$) is a global asymptotic attractor for $a < 0$, a center for $a = 0$, and a repellor for $a > 0$.

The function $V$, above, is an example of a Lyapunov function. And, in fact, we have just used Lyapunov's Direct Method, so-called because we were able to determine the stability of equilibrium directly, that is, without having to solve (6.12). The more general result is given in the following:

[119] Huntley and Johnson (1983, p. 117).

**Theorem 3** (Lyapunov). *Let $\bar{x}$ be a fixed point of $\dot{x} = f(x)$, $x \in \mathcal{R}^2$ and let $V : W \subset \mathcal{R}^2 \to \mathcal{R}$ be differentiable on some neighborhood $W$ of $\bar{x}$ and satisfy:*

$$
\begin{aligned}
&(i) \quad V(\bar{x}) = 0;\\
&(ii) \quad V(x) > 0 \text{ if } x \neq \bar{x};\\
&(iii) \quad \dot{V}(x) \leq 0 \text{ on } W - \{\bar{x}\}.
\end{aligned}
$$

*Then $\bar{x}$ is stable. If $\dot{V} < 0$ in (iii), then $\bar{x}$ is asymptotically stable.*[120]

In the previous example, $V = x_1^2 + x_2^2$. Quadratic forms ($V = ax_1^2 + bx_1x_2 + cx_2^2$) are often good candidates. While the method is very powerful, there is no formula for constructing Lyapunov functions; this is a bit of an art.

The typical Lyapunov function is a bowl-like surface whose level sets lie in a subset of the plane. Geometrically, the theorem says simply that asymptotic stability requires trajectories to cross these level sets in the *inward* direction, as shown in figure 6.4.

FIGURE 6.4 Asymptotic Stability

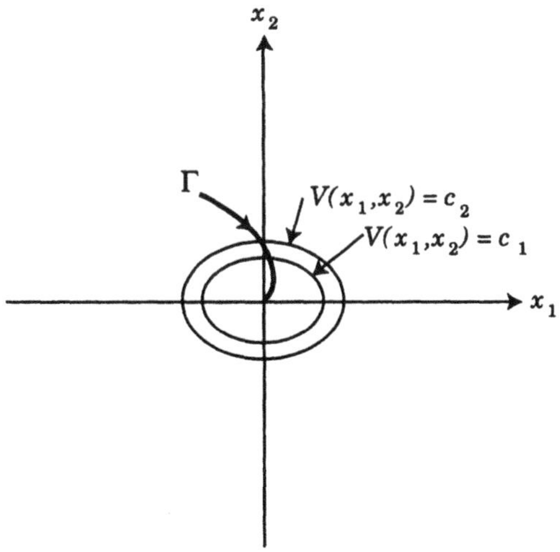


[120]Hirsch and Smale (1974, p. 193).

FIGURE 6.5 Close Up

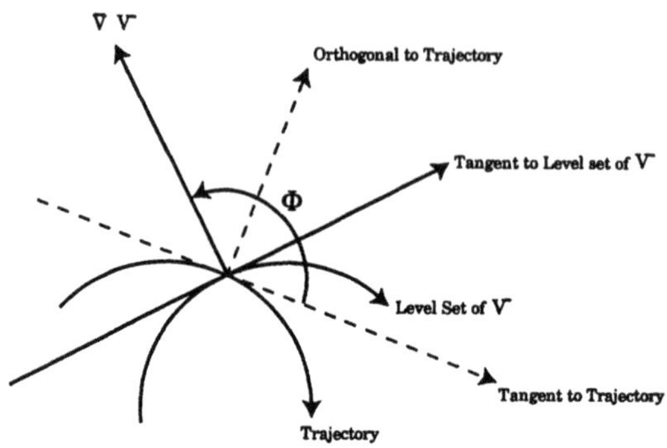


This seems eminently reasonable. Can we get from there to a more formal argument?[121] Let us "zoom in" on a point where this inward crossing takes place, as shown in figure 6.5.

Let $\Phi$ be the angle between $\nabla V$(*normal* to the level set of $V$) and the trajectory's tangent, $(\dot{x}_1, \dot{x}_2)$. Recall that for two vectors $a$ and $b$, $a \cdot b = ||a|| \cdot ||b|| \cos\theta$, where $\theta$ is the angle between $a$ and $b$. As long as the flow points inward, we must have $\pi/2 < \Phi < 3\pi/2$, which implies $\cos\Phi < 0$. But since $||\nabla V||$ and $||(\dot{x}_1, \dot{x}_2)||$ are positive, $\nabla V \cdot (\dot{x}_1, \dot{x}_2) < 0$, but $\nabla V \cdot (\dot{x}_1, \dot{x}_2) = \dot{V}$, and we are through.

As a second example of Lyapunov's direct method, let us establish a nice general property of gradient systems, which are important in many applications. By way of definition, let $U(x)$ be a real-valued differentiable function on $\mathcal{R}^n$. Then

$$\dot{x} = -\nabla U(x) \tag{6.17}$$

is a *gradient system*. For every $i$, $dx_i/dt = -\partial U/\partial x_i$; the velocity of $x_i$ equals the negative partial of $U$ with respect to $x_i$. For physicists, $U$ is most naturally interpreted as a potential of some sort. But, one can think of (6.17) as stipulating a rule of adjustment for each $x_i$. For example, in the backpropagation neural network, connection weights are adjusted in proportion to an error gradient.[122] In evolutionary game theory, phenotypic frequencies are adjusted (by selection) in the direction of fitness gradients.[123] Gradient systems have the following:

**Property.** *A gradient system is asymptotically stable at* $\bar{x}$ *if* $U$ *has an isolated local minimum there.*

[121] For a rigorous proof, see Hirsch and Smale (1974, p. 194).
[122] Rumelhart and McClelland (1986), Freeman and Skapura (1991).
[123] Hofbauer and Sigmund (1988).

**Proof.** We demonstrate that $V = U(x) - U(\bar{x})$ is a Lyapunov function. We need only establish the properties given in Theorem 3. First, $V(\bar{x}) = 0$ by construction. Now, to call $\bar{x}$ an isolated minimum is to assert the existence of some neighborhood $W$ of $\bar{x}$ in which $V(x) > 0$ for $x \neq \bar{x}$, which is the second Lyapunov property. Third, we show that $\dot{V}(x) < 0$ on $W - \{\bar{x}\}$. Computing,

$$\begin{aligned}
\dot{V} &= \sum_{i=1}^{n} \frac{\partial U}{\partial x_i}\frac{dx_i}{dt} && \text{(by the Chain Rule)} \\
&= \sum_{i=1}^{n} \frac{\partial U}{\partial x_i}\left(-\frac{\partial U}{\partial x_i}\right) && \text{(since gradient)} \\
&= -\sum_{i=1}^{n} \left(\frac{\partial U}{\partial x_i}\right)^2 \\
&= -||\nabla U(x)||^2 < 0 \text{ on } W - \{\bar{x}\}\,. && \square
\end{aligned}$$

Since they are related to Lyapunov functions and arise in some of the other lectures, I will briefly discuss Hamiltonian flows. A planar dynamical system $\dot{x} = f(x)$ is said to admit an autonomous Hamiltonian formulation if there exists a $C^1$ function $H : \mathcal{R}^2 \rightarrow \mathcal{R}$ such that

$$\begin{aligned}
\dot{x}_1 &= -\frac{\partial H}{\partial x_2} = f_1(x_1, x_2)\,, \\
\dot{x}_2 &= \frac{\partial H}{\partial x_1} = f_2(x_1, x_2)\,.
\end{aligned} \tag{6.18}$$

In that case, $H$ is said to be a *Hamiltonian* of the system.[124] As a general proposition, we can consider $H(x(t))$ along trajectories of (6.18). We differentiate $H$ with respect to time exactly as we did the Lyapunov function above:

$$\begin{aligned}
\dot{H}(x(t)) &= \frac{\partial H}{\partial x_1}\dot{x}_1 + \frac{\partial H}{\partial x_2}\dot{x}_2 \\
&= \frac{\partial H}{\partial x_1}\left(-\frac{\partial H}{\partial x_2}\right) + \frac{\partial H}{\partial x_2}\frac{\partial H}{\partial x_1} = 0\,.
\end{aligned} \tag{6.19}$$

Hence $H$ is constant, or conserved, along trajectories of (6.18). Total mechanical energy is the archetypal Hamiltonian from classical physics. Notice that, with $F = (f_1, f_2)$, $\nabla H \cdot F = 0$, whereas in the gradient case $\nabla U \cdot F < 0$. For this reason, we speak of gradient fields as *dissipative* in contrast to *conservative* Hamiltonian fields.

[124] For the definition of $n$ dimensional Hamiltonians, and the interesting property that $n$ must be even, see Jackson (1989, vol. 1, p. 20).

For another perspective, recall that the divergence of a vector field can be interpreted as the rate of expansion per unit volume of a fluid whose flow is modeled by $F$. In the event $F$ is Hamiltonian, we find:

$$\begin{aligned}
\operatorname{div} F &= \nabla \cdot F \\
&= \frac{\partial f_1}{\partial x_1} + \frac{\partial f_2}{\partial x_2} \\
&= -\frac{\partial^2 H}{\partial x_1 \partial x_2} + \frac{\partial^2 H}{\partial x_2 \partial x_1} \\
&= 0\,,
\end{aligned}$$

since $H$ is $C^1$ and so the mixed partials are equal. Hamiltonian flows are volume preserving, a result sometimes known as *Liouville's Theorem.*[125]

Now, it is evident from the above considerations that Hamiltonian flows cannot have sinks or sources as equilibria since constancy along trajectories would then clearly be violated. Centers, by contrast, are admissible, and, less obviously, so are (certain) saddles.[126]

Returning, then, to the issue of local stability analysis, suppose we encounter a system we know to be Hamiltonian (whether or not we can display $H$) and we have an equilibrium where the eigenvalues are purely imaginary. The equilibrium is nonhyperbolic, so linearization fails, but we can instantly conclude that it is a center. Why? From Poincaré's Theorem above, we know it is a center or a focus. And, since the system is Hamiltonian, we know it is a center or a saddle. Hence, it is a center. Nonlinear systems exhibit behaviors quite unlike those we have discussed to this point.

## LIMIT CYCLES

To introduce the central, and (for autonomous planar systems) uniquely nonlinear, phenomenon of the *limit cycle* and some of the associated theory, consider the following system—a variation on (6.12).

$$\begin{aligned}
\dot{x}_1 &= x_2 + x_1(\lambda - x_1^2 - x_2^2)\,, \\
\dot{x}_2 &= -x_1 + x_2(\lambda - x_1^2 - x_2^2)\,.
\end{aligned} \tag{6.20}$$

In polar coordinates, this takes the form:

$$\begin{aligned}
\dot{\theta} &= -1\,, \\
\dot{r} &= r(\lambda - r^2)\,.
\end{aligned} \tag{6.21}$$

[125] Guckenheimer and Holmes (1983, p. 47).

[126] See Jackson (1989, vol. 1, p. 237).

For $\lambda \leq 0, \dot{r} < 0$ and solutions spiral to the origin as $t \to \infty$. But, if $\lambda > 0$, there are three cases to consider:

$$\text{If } r^2 > \lambda\,, \quad \dot{r} < 0\,.$$
$$\text{If } r^2 < \lambda\,, \quad \dot{r} > 0\,.$$
$$\text{If } r^2 = \lambda\,, \quad \dot{r} = 0\,.$$

This tells us that trajectories beginning outside the circle, $r^2 = \lambda$, wind inward while trajectories (the origin aside) beginning inside that circle wind outward, and that as $t \to \infty$, all these trajectories spiral toward the circle $r^2 = \lambda$, itself a periodic orbit of (6.21). Because it is, in this sense, an attractor as $t \to \infty$ and a cycle, the orbit $r = \sqrt{\lambda}$ is called a *stable limit cycle.* For the time reversed system, the same object is an unstable limit cycle, for obvious reasons.

While the more modern technical definition (in terms of $\omega$-limit sets) involves further apparatus, Minorsky's makes immediately clear the distinction between limit cycles and the orbits surrounding neutrally stable centers. (I italicize the relevant phrase.)

> "A limit cycle is a closed trajectory (hence the trajectory of a periodic solution) such that *no trajectory sufficiently near it is also closed.* In other words, a limit cycle is an isolated closed trajectory. Every trajectory beginning sufficiently near a limit cycle approaches it either for $t \to \infty$ or for $t \to -\infty$, that is, it either winds itself upon the limit cycle, or unwinds from it. If all nearby trajectories approach a limit cycle $C$ as $t \to \infty$, we say that $C$ is *stable*; if they approach $C$ as $t \to -\infty$ we say that $C$ is *unstable.* If the trajectories on one side of $C$ approach it while those on the other depart from it, we sometimes say that $C$ is *semi-stable* although from a practical point of view $C$ must be considered unstable" (Minorsky, 1962, pp. 71–72).

The stable limit cycle is the basic model for all self-sustained oscillators—those which return, or recover, to some fundamental periodic orbit when perturbed from it. As Hirsch and Smale put it, "for a periodic solution to be viable in applied mathematics, this or some related stability property must be satisfied."[127] The stable oscillations, "beating" of the human heart (which returns to some normal rate after we raise it by sprinting), cycles of predator-prey systems, and various electrical circuits are three among myriad examples. Business cycles and certain periodic outbreaks of social unrest (see lecture 4) are, quite possibly, others.

[127] Hirsch and Smale (1974, p. 211).

## HILBERT'S 16TH PROBLEM

Quite aside from their practical importance, limit cycles also occupy a venerable position in the history of twentieth-century mathematics. In 1900, at the Second International Congress of Mathematicians in Paris, David Hilbert presented his famous 23 problems. The second part of Hilbert's 16th problem may be stated as follows: Determine the maximum number and position of limit cycles for the system

$$\begin{aligned} \dot{x} &= f(x,y)\,, \\ \dot{y} &= g(x,y)\,, \end{aligned} \tag{6.22}$$

where $f$ and $g$ are $n$th-degree polynomials

$$P_n \equiv \sum_{i+j=0}^{n} a_{ij}x^i y^j \equiv a_{00} + a_{10}x + a_{01}y + a_{20}x^2 + a_{11}xy + \ldots + a_{0n}y^n\,.$$

Defining the Hilbert numbers

$$H(n) = \ \sup_{f,g\in P_n}\{\text{number of limit cycles of (6.22)}\}\,,$$

the problem is to determine $H(n)$ for arbitrary $n$. It is not hard to establish that $H(0) = H(1) = 0$. But, for $n \geq 2$, we know precious little. Il'yashenko (1991) has shown that $H(n)$ is finite. We also know that $H(2) \geq 4$ and that $H(3) \geq 6$. And, after nearly a century, that's about it![128]

Small wonder that in 1947, Minorsky could write, "Perhaps it is not too great an exaggeration to say that the principal line of endeavor of nonlinear mechanics at present is a search for limit cycles."[129] In today's world of chaotic dynamics and strange attractors, this would be an exaggeration, but limit cycles remain eminently worthy quarry and—by way of Poincaré maps—they lead to the study of discrete dynamical systems, where low-dimensional chaos can be found. (In autonomous differentiable systems, chaos only arises in dimensions three or greater.)

The main theoretical tool in the search for limit cycles is the Poincaré-Bendixson Theorem, whose contemporary statement requires us to define $\omega$ (omega) limit points.

The basic idea of $\omega$ and $\alpha$ limit points is simple. Any point to which a trajectory converges in forward time is an $\omega$ limit point and any point to which the time-reversed trajectory converges is an $\alpha$ limit point. The technical definition is just a bit more discriminating. Let $\Gamma(t) = (x(t), y(t))$ be a trajectory of

$$\begin{aligned} \dot{x} &= f(x,y)\,, \\ \dot{y} &= g(x,y);\, f,g \in C^1\,. \end{aligned} \tag{6.23}$$

[128] Il'yashenko (1991); Coleman (1978).

[129] Cited in Jackson (1989, vol. 1, p. 234).

**Definition.** A point $P \in \mathcal{R}^2$ is an $\omega$ limit point of $\Gamma$ if there exists a sequence $\{t_n\}$ such that $t_n \to \infty$ as $n \to \infty$ and

$$\lim_{n\to\infty} \Gamma(t_n) = P\,.$$

The set of all such $P$s is called an $\omega$ limit set; $\alpha$ limit points and limit sets are analogously defined, with $t_n$ approaching $-\infty$ rather than $+\infty$.

A limit cycle, then, is rigorously defined as *a periodic orbit that is the $\omega$ or $\alpha$ limit set of other orbits.*[130] Denoting the $\omega$ limit set of a trajectory $\Gamma$ by $\omega(\Gamma)$, we state the celebrated Poincaré-Bendixson Theorem, perhaps the centerpiece of nonlinear dynamics in the plane.[131]

**Theorem 4** (Poincaré-Bendixson). *Let $\Gamma(t) = (x(t), y(t))$ be a trajectory of (6.23) such that, for $t \geq t_0$, $\Gamma(t)$ remains in a closed and bounded region of the plane containing no equilibrium points. Then* either *$\Gamma$ or $\omega(\Gamma)$ is a periodic orbit.*

The proof rests on the fundamental Jordan Curve Theorem: a simple closed curve divides the plane into two connected regions and is their common boundary; one region (called "the inside") is bounded and the other (called "the outside") is unbounded. On the face of it, nothing could appear more obvious. Yet, the proof is hard.[132] In other words, it is hard to really put the basic concepts of "inside" and "outside" on a firm footing. How many other "obvious" things must we not understand?

The main difficulty in applying the Poincaré-Bendixson Theorem lies in establishing an equilibrium-free closed and bounded "trapping" region. Sometimes it is easy, as in the system (6.20) above. In this case, we first imagine a circle centered at the origin with radius $r_0 < \sqrt{\lambda}$. Since $\dot{r} > 0$, *all* trajectories cross this circle *outward.* For a circle of radius $r_1 > \sqrt{\lambda}$, $\dot{r} < 0$ and *all* trajectories cross *inward.* Hence, every trajectory that, at $t = t_0$, is inside the closed and bounded region between these circles, the annulus $r_0 \leq r \leq r_1$, remains there for all $t > t_0$. The annulus contains no equilibria of (6.20). Hence, by the Poincaré-Bendixson Theorem, it must contain a periodic orbit.

In cases of this sort, the theorem makes good intuitive sense. A trajectory starting inside the circle $r = \sqrt{\lambda}$ is spiraling out. It cannot intersect itself (by uniqueness) and it can't "escape"; so it winds out to a periodic orbit.

As I said, establishing a compact[133] trapping region can be hard. The other limitation is that the Poincaré-Bendixson Theorem is false for autonomous systems

[130] To be completely rigorous, we should carefully define the "distance" from a tra ectory to a *set*, as in Waltman (1986, pp. 137–38 and 142).

[131] This statement parallels Waltman (1986, pp. 143–44).

[132] See Gamelin and Greene (1980).

[133] An arbitrary metric space $A$ is *compact* if and only if every open cover has a finite subcover. The equivalence to closed and bounded *in* $\mathcal{R}^n$ is a theorem (Heine-Borel). See Marsden (1974).

of dimension greater than 2 (basically because we lose the Jordan Curve Theorem). There is another powerful theorem concerning limit cycles that applies in higher dimensions. The theorem also concerns *bifurcation*, a concept central to the study of dynamical systems and chaos.

## BIFURCATION

As the simplest introduction to this deep area, begin with a vector field of the sort we have encountered: $\dot{x} = f(x, \mu)$, $x \in \mathcal{R}^2$ and $\mu$ is a single real parameter. When we vary $\mu$ continuously, how does the overall flow in phase space, $x_\mu(t)$, behave? And, specifically, are there points—$\mu$ values—at which the basic, topological, structure of the phase portrait changes? At such points—the so-called bifurcation points—the field $f(x, \mu)$ is "structurally unstable" in the sense that "nearby fields" $f(x, \mu + \varepsilon)$ have a different topological structure. The phase portrait *at* $\mu$, in other words, is not topologically equivalent to the phase portrait *at* $\mu + \varepsilon$. In turn, and now more precisely, a field $f$ is said to be "structurally stable" *if and only if there is a neighborhood (in the space of fields) of f such that all fields in the neighborhood are topologically equivalent to f*. Recall that two fields are topologically equivalent if and only if there is a homeomorphism mapping the orbits of one onto those of the other and preserving direction in time. (Under this interpretation, "phase portraits" really identify topological equivalence classes of fields). So, in a family of differential equations

$$\dot{x} = f(x, \mu), \;\; x \in \mathcal{R}^n, \;\; \mu \in \mathcal{R}^k, \tag{6.24}$$

we take the following[134]

**Definition.** A value $\mu_0$ for which the flow of (6.24) is not structurally stable is a bifurcation value of $\mu$.

Now, to really discuss structural stability, we must nail down the term "neighborhood of a field," and this requires that we impose a metric structure on the relevant set of vector fields. This can rigorously be done in a general metric space context, but will not be done here. Smale's essay, "What is Global Analysis?"[135] remains a wonderful starting point, while Hale and Koçak get to a "big" theorem (on genericity) about as quickly as possible.[136]

Keeping things intuitive, it is hardly debatable that if, for some $\mu$, the phase flow has a single point attractor but, for even slightly larger $\mu$, it has a stable limit

[134] Guckenheimer and Holmes (1983, p. 119).
[135] Smale (1980, pp. 84–89).
[136] Hale and Koçak (1991, p. 393).

cycle, then a bifurcation occurs at $\mu$. And this is the behavior considered in the Hopf Bifurcation Theorem.[137]

**Theorem 5** (Hopf Bifurcation Theorem in $\mathcal{R}^2$). *Suppose the parametrized system $\dot{x} = f(x, \mu)$, $x \in \mathcal{R}^2$, $\mu \in \mathcal{R}$ has a fixed point at the origin for all values of the real parameter $\mu$. Further, suppose the eigenvalues $\lambda_1(\mu)$ and $\lambda_2(\mu)$ of the ($\mu$-dependent) Jacobian of $f$, at zero, are purely imaginary for $\mu = \mu^*$. If the real part of the eigenvalues, $Re\lambda_1(\mu)(= Re\lambda_2(\mu)$ since $\lambda_1 = \bar{\lambda}_2)$, satisfies*

$$\frac{d}{d\mu}(\mathrm{Re}\lambda_1(\mu))|_{\mu=\mu^*} > 0$$

*and the origin is asymptotically stable when $\mu = \mu^*$, then:*

(a) *$\mu = \mu^*$ is a bifurcation point of the system;*

(b) *for $\mu \in (\mu_1, \mu^*)$ some $\mu_1 < \mu^*$, the origin is a stable focus;*

(c) *for $\mu \in (\mu^*, \mu_2)$ some $\mu_2 > \mu^*$, the origin is an unstable focus surrounded by a stable limit cycle whose size increases with $\mu$.*

Despite the theorem's forbidding appearance, it is explicable without too much agony. First, the theorem posits complex conjugate eigenvalues that are purely imaginary at the bifurcation point $\mu = \mu^*$, which is named in part (a). Of course $\mu^*$ is named that for a reason; the dynamics undergo a notable change at that point. We have that

$$\mathrm{Re}\lambda_1(\mu^*) = 0 \text{ but } \frac{d}{d\mu}(\mathrm{Re}\lambda_1(\mu^*)) > 0\,.$$

That is to say, at $\mu^*$, the real part of the eigenvalue is zero, but its rate of change, the slope, is positive; so, the value must be negative a little to the left, and positive a little to the right, of $\mu^*$. On either side, then, the equilibrium is hyperbolic and so, by linearized stability analysis, we have the stable and unstable foci predicted in (b) and (c). Overall, then, we would expect a change of stability as $\mathrm{Re}\lambda_1(\mu)$ "crosses the imaginary axis." It is, however, the birth of a stable limit cycle in particular that is surprising, and harder to prove. Indeed a big-league proof requires material (center manifold and normal form theory) beyond the scope of this lecture. But, there is a "physical" way to think about it. Compare the earlier dynamical system (6.12) with the Hopf case. In the first, the origin is *not* asymptotically stable and the unstable focus (to the right of the critical point in this case) simply spirals out; no limit cycle takes shape. But, in the second, Hopf, case, where a limit cycle *does* form, the origin *is* asymptotically stable. Its "force of attraction," if you will, while too weak to block the passage of Re($\lambda$) across the imaginary axis (it has

[137]This statement parallels Arrowsmith and Place (1990, p. 205). See also Marsden and McCracken (1976), and Guckenheimer and Holmes (1983, pp. 151–52).

"momentum" $\text{Re}\lambda'(\mu^*) > 0$ recall), is sufficient to prevent an unbounded escape of the orbit and so (by uniqueness again, as in Poincaré-Bendixson) a limit cycle emerges. The theorem is easy to use.

HOPF EXAMPLE #1. Consider the dynamical system

$$\dot{x}_1 = x_2 + x_1(\mu - x_1^2 - x_2^2),$$
$$\dot{x}_2 = -x_1 + x_2(\mu - x_1^2 - x_2^2).$$

The Jacobian at $\mu$ is

$$DF(x,\mu) = \begin{pmatrix} \mu - 3x_1^2 - x_2^2 & 1 - 2x_1x_2 \\ -1 - 2x_2x_1 & \mu - 2x_1x_2 - 2x_2^2 \end{pmatrix}.$$

Evaluated at the equilibrium $\bar{x} = 0$, the $\mu$-dependent Jacobian is

$$DF(0,\mu) = \begin{pmatrix} \mu & 1 \\ -1 & \mu \end{pmatrix}.$$

The characteristic equation is

$$P(\lambda) = (u - \lambda)^2 + 1 = 0,$$

yielding eigenvalues

$$\lambda_{1,2} = \mu \pm i.$$

Let us now check if the Hopf bifurcation conditions are met.

First, the eigenvalues of the Jacobian at zero are purely imaginary if $\mu = 0$. So, define $\mu^* = 0$. Second, since $\text{Re}\lambda(\mu) = \mu$, its derivative with respect to $\mu$ is 1, so

$$\frac{d}{d\mu}(\text{Re}\lambda(\mu))|_{\mu=\mu^*} = 1 > 0.$$

Finally, we need to check that, for $\mu = \mu^* = 0$, the origin, $\bar{x} = 0$, is asymptotically stable. Since the eigenvalues at $\mu = \mu^*$ are purely imaginary, the equilibrium is nonhyperbolic and linearized stability analysis fails. Lyapunov's direct method works well, however. Define

$$V = \frac{x_1^2 + x_2^2}{2}.$$

Then, with $r^2 = x_1^2 + x_2^2$, we obtain

$$\dot{V} = x_1\dot{x}_1 + x_2\dot{x}_2 = r^2(\mu - r^2).$$

And, at $\mu = \mu^* = 0$ we have

$$\dot{V} = -r^4 < 0 \text{ on } R - \{0\}.$$

Hence, by the Hopf Bifurcation Theorem, $\mu = \mu^*$ is a bifurcation point and for some $\mu > \mu^*$, the origin is an unstable focus surrounded by a limit cycle whose size grows with $\mu$, as shown in figure 6.6.

FIGURE 6.6 Bifurcation to a Limit Cycle.

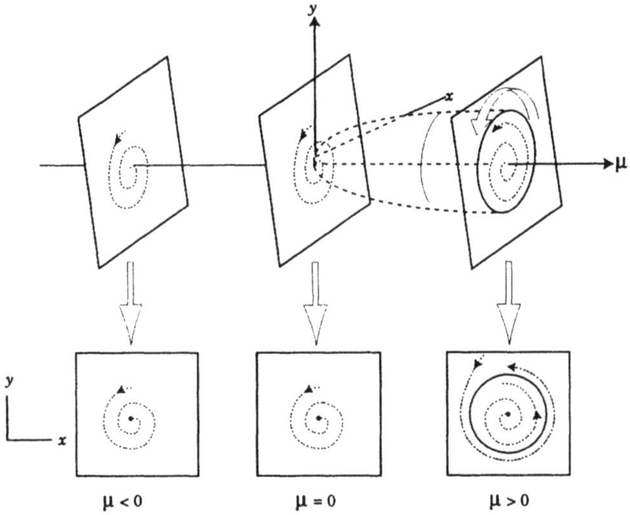


Source: Based on Wiggins (1990, p. 274).

HOPF EXAMPLE # 2 (The Van der Pol oscillator). Another, very famous, example is the Van der Pol oscillator

$$\ddot{y} - (2\mu - y^2)\dot{y} + y = 0\,,$$

or, equivalently,

$$\begin{aligned}\dot{x}_1 &= x_2\,,\\ \dot{x}_2 &= -x_1 + 2\mu x_2 - x_1^2 x_2\,.\end{aligned}$$

The Jacobian is

$$DF(x,\mu) = \begin{pmatrix} 0 & 1 \\ -1 - 2x_1x_2 & 2\mu - x_1^2 \end{pmatrix}.$$

At $\bar{x} = 0$, we have

$$DF(0,\mu) = \begin{pmatrix} 0 & 1 \\ -1 & 2\mu \end{pmatrix},$$

with characteristic equation

$$P(\lambda) = \lambda^2 - 2\mu\lambda + 1 = 0\,.$$

The eigenvalues are $\lambda_{1,2} = \mu \pm i\sqrt{1-\mu^2}$. Regarding the Hopf bifurcation requirements, the eigenvalues of the Jacobian at $\bar{x} = 0$ are again purely imaginary if $\mu = 0$. So define $\mu^* = 0$. $\text{Re}\lambda(\mu) = \mu$ once more, so we again have

$$\frac{d}{d\mu}(\text{Re}\lambda(\mu))|_{\mu=\mu^*} = 1 > 0\,.$$

Finally, at $\mu = \mu^*$, the bifurcation point, the origin is nonhyperbolic, so linearization fails. But, once more, it's Lyapunov to the rescue! Taking

$$V = \frac{x_1^2 + x_2^2}{2}\,,$$

we have

$$\dot{V} = x_1\dot{x}_1 + x_2\dot{x}_2 = 2\mu x_2^2 - x_1^2 x_2^2\,.$$

And, at $\mu = \mu^* = 0$,

$$\dot{V} = -x_1^2 x_2^2 < 0 \text{ on } \mathcal{R}^2 - \{0\}\,.$$

Hence, $\bar{x}$ is asymptotically stable. Just so you won't think all limit cycles are circular, the Van der Pol oscillator is shown in figure 6.7.

FIGURE 6.7 Van der Pol Oscillator

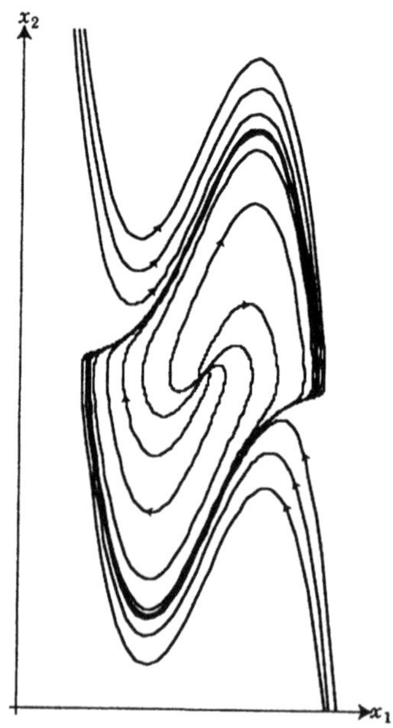

For the Van der Pol equation and its variants (in the family of so-called Liénard equations), the actual construction of a compact invariant (trapping) set containing no equilibria, as called for by Poincaré-Bendixson, is fairly arduous. The Hopf bifurcation theorem demonstrates the existence of a stable limit cycle quite painlessly (though, unlike an explicit construction, it says little about its shape).

Now, the examples above are all cases of stable limit cycles in $\mathcal{R}^2$; orbits wind toward them—they are attractors. There is a very ingenious way to represent these orbits as fixed points of discrete maps in a lower dimensional space. The method, like so much else, is due to Poincaré, and bears his name.

## POINCARÉ MAPS

The basic idea is easily conveyed in the plane. For a sophisticated treatment see Wiggins (1990). For illustrative purposes, imagine a circular stable limit cycle centered at the origin surrounding an unstable focus, as shown in figure 6.8.

FIGURE 6.8 Crossings

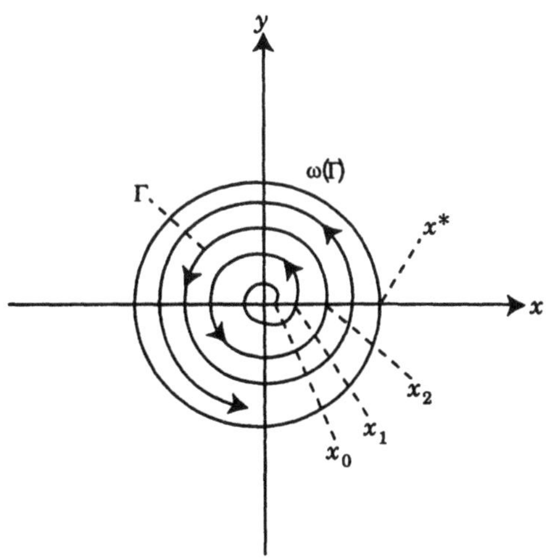


Given some initial point $x_0$ on, say, the positive $x$-axis, we can follow the trajectory around until it crosses the $x$-axis again, say, at $x_1$. We call $x_1$ the point of first return of $x_0$. Then, $x_2$ is the first return of $x_1$, and so forth. For a given cross section $\Sigma$ (these are called Poincaré sections), here the positive $x$-axis, this return map, which we denote $\Pi$, is called the *Poincaré map*.[138] The idea is to reduce the study of continuous time flows in two dimensions to the study of associated discrete time systems (maps) in one dimension. Very elegant simplifications result.

For example, the trajectory through a point $x^*$ is a periodic orbit if and only if $x^*$ returns to itself under the Poincaré map; that is to say, $\Pi(x^*) = x^*$. Demonstrating the existence of a limit cycle or other periodic orbit (a continuous curve) thus reduces to exhibiting a fixed *point* of the discrete Poincaré map. In turn, a limit cycle is stable if the fixed point of the Poincaré map is stable. More formally, a periodic orbit $\Gamma$ with $x^* \in \Gamma$ is asymptotically stable if $x^*$ is an asymptotically stable fixed point of $\Pi$—that is, if $\Pi'(x^*) < 1$. If $\Pi'(x^*) > 1$, $\Gamma$ is unstable.[139]

Conversely, if $\Gamma$ is stable, then $\Pi'(x^*) \leq 1$. To see this, imagine a stable limit cycle whose Poincaré map has the fixed point $x^*$. Trajectories beginning outside the limit cycle $\omega(\Gamma)$ are winding down onto it. By the uniqueness of orbits, the crossings $x_n$ are approaching $x^*$ monotonically from above. Thus,

$$\begin{aligned} &x^* < x_{n+1} < x_n\,, \qquad \text{and} \\ &0 < x_{n+1} - x^* < x_n - x^*\,. \end{aligned}$$

Since $x_{n+1} = \Pi(x_n)$ and $x^* = \Pi(x^*)$, we have

$$\begin{aligned} &\Pi(x_n) - \Pi(x^*) < x_n - x^*\,, \\ &\frac{\Pi(x_n) - \Pi(x^*)}{x_n - x^*} < 1\,, \qquad \text{and} \\ \lim_{n\to\infty} &\left(\frac{\Pi(x_n) - \Pi(x^*)}{x_n - x^*}\right) = \Pi'(x^*) \leq 1\,, \end{aligned} \tag{6.25}$$

as was to be shown.[140]

The Poincaré map connects the world of continuous flows in some dimension to the world of discrete maps in a space one dimension lower. The lower dimensional entity "talks about" the higher dimensional one. A sort of "inverse problem" is worth mentioning. Given any discrete map, of which continuous flow (or flows) is it the Poincaré map? The theory of suspensions treats this.[141]

Well, what does one of these Poincaré maps actually look like? We take an example from Guckenheimer and Holmes.[142] Let the dynamical system in $\mathcal{R}^2$ be

$$\begin{aligned} \dot{x} &= x - y - x(x^2 + y^2)\,, \\ \dot{y} &= x + y - y(x^2 + y^2)\,. \end{aligned}$$

[138] A subtlety we will ignore is that the map may depend on the section. See Wiggins (1990).

[139] Hale and Koçak (1991, p. 376).

[140] I thank Jean-Pierre Langlois for helping me tidy this argument.

[141] See Arrowsmith and Place (1990, Section 1.7) and Jackson (1989, vol. 1, Section 4.10).

[142] Guckenheimer and Holmes (1983, p. 23).

As the Poincaré section, take the positive $x$-axis:

$$\Sigma = \{(x, y) \in \mathcal{R}^2 | x > 0, y = 0\} .$$

Transforming to polar coordinates, we obtain

$$\begin{aligned} \dot{r} &= r(1 - r) , \\ \dot{\theta} &= 1 , \end{aligned} \tag{6.26}$$

and the section becomes

$$\Sigma = \{(r, \theta) \in \mathcal{R}^1 \times S^1 | r > 0, \theta = 0\} .$$

A product system, (6.26) is easily solved by elementary methods for the flow $x_t(r_0, \theta_0)$:

$$x_t(r_0, \theta_0) = \left[ \left( 1 + \left( \frac{1}{r_0^2} - 1 \right) e^{-2t} \right)^{-1/2}, t + \theta_0 \right] .$$

Since $\dot{\theta} = 1$, it is clear (separating variables) that the period equals $2\pi$ so that the Poincaré map is

$$\Pi(r_0) = \left[ 1 + \left( \frac{1}{r_0^2} - 1 \right) e^{-4\pi} \right]^{-1/2} .$$

We have $\Pi(1) = 1$, a fixed point, so there is a limit cycle of radius 1. And, since $\Pi'(1) = e^{-4\pi} < 1$, it is stable.

Despite the tools we have developed—the use of polar coordinates, the Poincaré-Bendixson and Hopf Bifurcation Theorems, and Poincaré Maps—finding limit cycles can be hard (consider Hilbert's 16th problem). Precluding their existence can be easier, even when the dynamical system looks, at first glance, very forbidding.

## NEGATIVE TESTS

For example, consider:

$$\begin{aligned} \dot{x} &= 2e^y \sin^7 y - y^{1/2} + x = f(x, y) , \\ \dot{y} &= xe^x \cos x^2 + y = g(x, y) . \end{aligned} \tag{6.27}$$

Does this dynamical system have any cycles in $\mathcal{R}^2$? Well, the presence of some trigonometric functions might embolden one to venture a tentative yes. But, at least to me, this specimen looks pretty inscrutable at first glance. Amazingly, there is a theorem that will let you decide the issue virtually at sight! It is called Bendixson's negative test. To state the theorem, let $f$ and $g$ define a vector field on a (simply connected) region $D \subset \mathcal{R}^2$. Call that field $\Omega = (f, g)$. Then, we state:

**Theorem 6** (Bendixson's Negative Test). *If div $\Omega$ has fixed sign in a region D, then $\Omega$ has no cycles in D.*

We will prove a more powerful result shortly but, first, look how simple this makes the problem (6.27) above. At sight, we obtain

$$\frac{\partial f}{\partial x} = \frac{\partial g}{\partial y} = 1, \text{ so div } \Omega \equiv \frac{\partial f}{\partial x} + \frac{\partial g}{\partial y} = 2 > 0$$

and, *voilà*, there are no cycles. System (6.27) is of the form:

$$\begin{aligned} \dot{x} &= \Psi_1(y) + \Phi_1(x, y)\,, \\ \dot{y} &= \Psi_2(x) + \Phi_2(x, y)\,. \end{aligned}$$

Since $\partial\Psi_1/\partial x = \partial\Psi_2/\partial y = 0$, the $\Psi$s contribute nothing to the divergence of the field and, thus, no matter how wild they may look, are irrelevant to the question of cycles. As I mentioned, we actually have a stronger result.

**Theorem 7** (Bendixson-Dulac). *Given the system*

$$\begin{aligned} \dot{x} &= f(x, y)\,, \\ \dot{y} &= g(x, y)\,, \end{aligned} \tag{6.28}$$

*where f and g are smooth ($C^1$) in a simply connected region D, let $B(x, y)$ be a smooth function in D such that*

$$\frac{\partial(Bf)}{\partial x} + \frac{\partial(Bg)}{\partial y} \tag{6.29}$$

*has fixed sign in D. Then (6.28) has no closed trajectories in D.*

**Proof.** Assume there is a closed trajectory, $\partial D$, which is the boundary of a simply connected region $D$. By Green's Theorem,

$$\int\int_D \left[\frac{\partial(Bf)}{\partial x} + \frac{\partial(Bg)}{\partial y}\right] dxdy = \oint_{\partial D} (Bf)dy - (Bg)dx\,. \tag{6.30}$$

The right-hand side is zero since, factoring out $B$ and expanding the differentials, we obtain

$$\oint_{\partial D} B(f\dot{y} - g\dot{x})dt = \oint_{\partial D} B(fg - gf)dt = 0\,.$$

So, if there is a closed trajectory $\partial D$, the right-hand side of (6.30) is zero. Suppose now that expression (6.29), the integrand of the double integral above, does not change sign. Without loss of generality, suppose the integrand is always positive on

$D$. Then, clearly, the double integral will be positive, too, and we will be forced to deduce that

$$0 < \int\int_D \left[\frac{\partial(Bf)}{\partial x} + \frac{\partial(Bg)}{\partial y}\right] dxdy = \oint_{\partial D} B(fg - gf)dt = 0\,,$$

a contradiction. □

As a simple application of Bendixson-Dulac, let us prove Ragozin's negative criterion[143] for interacting species, bearing in mind, as always, that "species" could be interpreted in myriad other ways (e.g., armies, chemical concentrations). This theorem applies to Kolmogorov-type dynamical systems. These have the form:

$$\begin{aligned} \dot{x} &= xf(x,y)\,, \\ \dot{y} &= yg(x,y)\,. \end{aligned} \tag{6.31}$$

In effect, $\ln x$ plays the role in (6.31) that $x$ plays in (6.1), since (6.31) implies $\dot{x}/x = f(x,y)$, but $\dot{x}/x$ is $d/dt(\ln x)$. It is the *per capita* growth rate $\dot{x}/x$ that is given by $f$. Now, species are said to be *self-regulating* if the per capita growth rate decreases as the species population increases. If

$$\frac{\partial f}{\partial x} < 0 \text{ and } \frac{\partial g}{\partial y} < 0\,, \tag{6.32}$$

then, of course, each species is self-regulating. With this background we state:

**Proposition** (Ragozin). *A Kolmogorov system in which each species is self-regulating has no cycles in the population quadrant* $(x, y > 0)$.

**Proof.** To employ Bendixson-Dulac, let

$$xf(x,y) = H(x,y) \text{ and } yg(x,y) = J(x,y)\,.$$

Now, we seek some real function $B(x,y)$ such that Bendixson-Dulac applies. Coming up with such functions is a bit of an art, like inventing Lyapunov functions. Following Hethcote,[144] we try a function of the form:

$$B(x,y) = (xy)^j\,.$$

Specifically, let $j = -1$. Then, after a little algebra, we find that

$$\frac{\partial(BH)}{\partial x} + \frac{\partial(BJ)}{\partial y} = \frac{1}{y}\frac{\partial f}{\partial x} + \frac{1}{x}\frac{\partial g}{\partial y}\,.$$

[143] See Borelli and Coleman (1987, p. 494).
[144] Hethcote (1976, p. 349).

Since $x$ and $y$ are assumed positive, so are $1/x$ and $1/y$. But, $\partial f/\partial x$ and $\partial g/\partial y$ are negative by hypothesis. Hence

$$\frac{\partial(BH)}{\partial x} + \frac{\partial(BJ)}{\partial y} < 0,$$

and we are through. □

Clearly, the same reasoning applies to Kolmogorov systems in which each species is everywhere self-amplifying, or autocatalytic:

$$\frac{\partial f}{\partial x} > 0 \text{ and } \frac{\partial g}{\partial y} > 0. \tag{6.33}$$

Then, the system (6.31) has no cycles in the population quadrant.

In summary, as a corollary to Bendixson-Dulac: *for Kolmogorov systems, if either (6.32) or (6.33) obtain, there can be no cycles.*

Notice that the generalized Lotka-Volterra equations of lecture 1 are of Kolmogorov type.

$$\dot{x}_1 = x_1(r_1 + a_{11}x_1 + a_{12}x_2),$$
$$\dot{x}_2 = x_2(r_2 + a_{21}x_1 + a_{22}x_2).$$

If $a_{11}$ and $a_{22}$ are negative, each species is self-regulating; if $a_{11}$ and $a_{22}$ are positive, each species is self-amplifying. In either event, there are no cycles. (In the predator-prey variant where cycles *do* occur, $a_{11} = a_{22} = 0$.)

There is a powerful theorem of Kolmogorov governing the stability of those systems that bear his name. These, recall, are

$$\dot{x} = xf(x, y),$$
$$\dot{y} = yg(x, y).$$

Following May (1974), we state the result as follows: Kolmogorov systems have *either* a stable equilibrium point or a stable limit cycle provided $f$ and $g$ are $C^1$ functions of $x$ and $y$ on $x, y \geq 0$ and the following relations hold[145]:

$$\text{(i)} \quad \frac{\partial f}{\partial y} < 0,$$

$$\text{(ii)} \quad x\frac{\partial f}{\partial x} + y\frac{\partial f}{\partial y} < 0,$$

$$\text{(iii)} \quad \frac{\partial g}{\partial y} < 0,$$

$$\text{(iv)} \quad x\frac{\partial g}{\partial x} + y\frac{\partial g}{\partial y} > 0,$$

$$\text{(v)} \quad f(0, 0) > 0.$$

[145] See May (1974, pp. 87–88). May adds that "The theorem also usually holds when certain of the above conditions are equalities (=) rather than inequalities (< or >). Such cases need to be dealt with on their merits, but can often be seen to be sensible limiting cases of more general predator-prey equations which do obey the above criteria" (1974, p. 88).

Also, there must exist quantities $A$, $B$, and $C$ such that

$$\begin{aligned} &\text{(vi)} \quad f(0,A) = 0 \text{ with } A > 0\,, \\ &\text{(vii)} \quad f(B,0) = 0 \text{ with } B > 0\,, \\ &\text{(viii)} \quad g(C,0) = 0 \text{ with } C > 0\,, \\ &\text{(ix)} \quad B > C\,. \end{aligned}$$

Placing the theorem in context, May writes,

> "In more biological terms, Kolmogorov's conditions are roughly that (i) for any given population size (as measured by numbers, biomass, etc.), the per capita rate of increase of the prey species is a decreasing function of the number of predators, and similarly (iii) the rate of increase of predators decreases with their population size. For any given ratio between the two species, (ii) the rate of increase of the prey is a decreasing function of population size, while conversely (iv) that of the predators is an increasing function. It is also required that (v) when both populations are small the prey have a positive rate of increase, and that (vi) there can be a predator population size sufficiently large to stop further prey increase, even when the prey are rare. Condition (vii) requires a critical prey population size $B$, beyond which they cannot increase even in the absence of predators (a resource or other self limitation), and (viii) requires a critical prey size $C$ that stops further increase in predators, even if they be rare; unless (ix) $B > C$, the system will collapse" (May, 1974, pp. 88–89).

For a given Kolmogorov system, then, the programme is clear. First, establish whether the system satisfies the theorem's hypotheses, an eventuality May accounts as quite likely. He writes, "What has been lacking in the literature is not the derivation of the above theorem, but rather the realization that it applies to essentially all the conventional models people use." If the system meets all the conditions, then it possesses *either* a stable equilibrium point or a stable limit cycle. Our standard linearized stability analysis at the equilibrium will then reveal whether that point is stable, "whereupon we have the complete global stability character of this system laid bare."[146]

## INDEX THEORY

We have seen that nonlinear vector fields may have multiple fixed points. And we have developed a theory of stability allowing us to classify hyperbolic and (to some

[146] May (1974, p. 89).

extent) nonhyperbolic fixed points as stable or unstable nodes, foci, saddles, and centers, and to rule out and (with some creativity) to detect limit cycles. But, the theory is completely local; we have developed no theory of how these entities can combine, or "fit together." For instance, could one ever encounter a limit cycle with a single focus and two saddles inside?

There is a beautiful and penetrating theory that allows us to answer many such questions. It is called *index theory* and originates, once more, with Poincaré.[147] Among other things, index theory reveals what amount to topological conservation laws—people even speak of the "conservation of topological charge"! It also reveals that, from a particular topological perspective, centers, nodes, and foci are equivalent!

To begin, we recall the planar autonomous system

$$\begin{aligned} \dot{x} &= f(x,y)\,, \\ \dot{y} &= g(x,y), f,g \in C^1\,. \end{aligned} \tag{6.34}$$

The vector field $(f, g)$ is tangent to the solution flow in phase space at each point $(x, y)$. Now, let $C$ be a simple closed curve (not necessarily a trajectory) that does not intersect any fixed points of (6.34). This last proviso ensures that, on $C$, $f$ and $g$ are never simultaneously zero, so that $f^2 + g^2 \neq 0$, a condition whose importance will soon be evident. Since the field $(f, g)$ is tangent to the flow for all $(x, y)$, it is tangent to the flow for all $(x, y)$ *on* $C$. So imagine basing a small arrow at some point of $C$, letting the arrow point in the direction $(f, g)$. Now, slide the arrow's base exactly once around $C$ counterclockwise, allowing the arrow to assume the (continuously changing) direction $(f, g)$ at each point. Since the terminal point *is* the initial point, the terminal direction of the arrow coincides with the initial direction and, thus, having returned to its initial position, the arrow must have rotated through $2\pi$ radians an integral number of times: that integer is the *index* of $C$. Obviously it depends on the field.

For example, if the field is everywhere flowing right to left, the arrow will not rotate at all and, for any closed curve $C$, the index will be zero. By contrast, if the curve $C$ is itself a simple closed trajectory of (6.34), then, at any point on $C$, the tangent to the flow *is* the tangent to $C$, so that the index is 1, as illustrated in figure 6.9, which shows a variety of singularities and their indices (in parentheses).

[147] Poincaré's index is defined in Poincaré (1881, 1882, 1885, and 1886). For an account, see Lefschetz (1977, pp. 195–96). For developments of index theory, see also Jackson (1989, vol. 1, pp. 243–50), Jordan and Smith (1987, pp. 65–74), Guckenheimer and Holmes (1983, pp. 50–53), and Arnol'd (1991, pp. 309–18).

FIGURE 6.9 Field Directions and Indices.

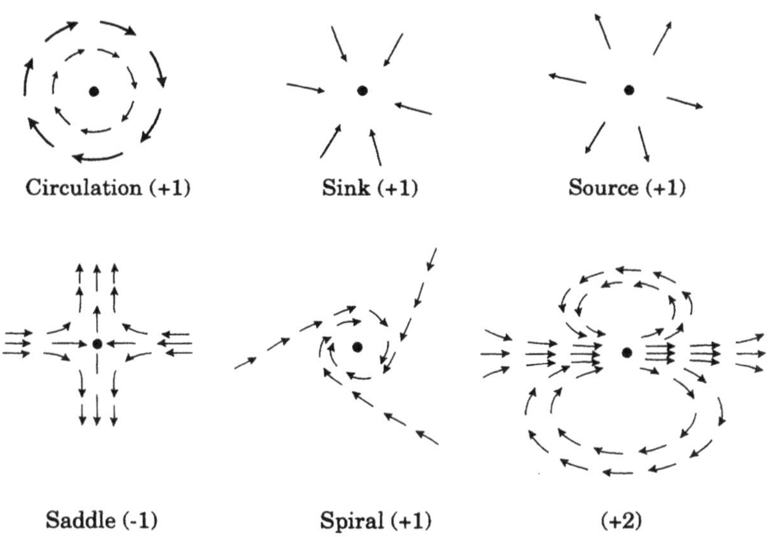


Source: Based on Guillemin and Pollack (1974, p. 133).

To put this on a more formal footing, we will assume that $f$ and $g$ and their first partials are continuous on the relevant sets—basically so there is no problem invoking Green's Theorem, which looms large here.

At any point $(x, y)$, the slope of the field vector is $dy/dx = g/f$. Hence, relative to the $x$-axis, the angle of inclination, $\Phi$, of the field vector must satisfy

$$\Phi = \arctan\left(\frac{g}{f}\right). \tag{6.35}$$

Defining $u = g/f$, we find

$$\begin{aligned} d\Phi &= \frac{\partial\Phi}{\partial u}\frac{\partial u}{\partial f}df + \frac{\partial\Phi}{\partial u}\frac{\partial u}{\partial g}dg \\ &= \frac{1}{1+u^2}\frac{\partial u}{\partial f}df + \frac{1}{1+u^2}\frac{\partial u}{\partial g}dg \\ &= \frac{f dg - g df}{f^2+g^2}, \end{aligned}$$

whence our concern that $f^2 + g^2 \neq 0$. Now, it is clear that the total change in $\Phi$ over one circuit around a simple closed curve $C$ is given by the line integral

$$\oint_C d\Phi .$$

But, as argued above, this is an integer multiple of $2\pi$, the integer being the index of $C$. Hence, denoting the index $I(C)$, we arrive at the relation

$$I(C) = \frac{1}{2\pi}\oint_C d\Phi = \frac{1}{2\pi}\oint_C \frac{fdg - gdf}{f^2 + g^2}. \tag{6.36}$$

From here, it is a short step to the next basic result:

**Theorem 8.** *Let $C$ be a simple closed curve that neither contains nor intersects any equilibrium points of (6.34). Then $I(C) = 0$.*

**Proof.** Let $R$ be the (simply connected) interior of $C$. Then (assuming $f$ and $g$ to be $C^1$) Green's Theorem yields

$$I(C) = \frac{1}{2\pi}\oint_{C=\partial R} \frac{fdg - gdf}{f^2+g^2} = \frac{1}{2\pi}\int\int_R \left[\frac{\partial}{\partial f}\left(\frac{f}{f^2+g^2}\right) + \frac{\partial}{\partial g}\left(\frac{g}{f^2+g^2}\right)\right] df dg.$$

But, differentiation will quickly show the integrand on the right to be zero. □

So, the index of a closed curve containing (and intersecting) no equilibria is zero. Now, what happens to the index if we deform $C$? If the index is invariant under (certain) deformations of $C$, then we really should not think of the index as characteristic of $C$ at all; but, then, what is the index really about? We begin to answer the question with a corollary to Theorem 8.

**Corollary 1.** *Let $C_1$ be a simple closed curve and let $C_2$ be a simple closed curve surrounding $C_1$. If there is no equilibrium of (6.34) on either curve or in the region between them, then $I(C_1) = I(C_2)$. Deformation has no effect.*

**Proof.** The proof is reminiscent of the standard complex variables proof of Cauchy's Theorem for multiply connected domains; we "cut" a multiply connected domain, leaving a simply connected one to which the more basic theorem applies. So, as shown in figure 6.10, let us cut the annulus between $C_1$ and $C_2$ at point $A$ with a segment $K$. Starting at $A$, we integrate counterclockwise around $C_1$, then down $K^-$, around $C_2$ (now clockwise, reversing sign), and out $K^+$ to the start. The closed[148] curve just traced neither intersects nor contains any equilibria, so its index is zero by Theorem 8. That is,

$$\frac{1}{2\pi}\left\{\oint_{C_1} d\Phi + \oint_{K^-} d\Phi - \oint_{C_2} d\Phi + \oint_{K^+} d\Phi\right\} = 0.$$

[148]Throughout, I will ask your indulgence for a slight mathematical indiscretion. Strictly speaking, if we cut the annulus with a line segment, then the path traced in the proof is not a *simple* closed curve since $K^+$ intersects $K^-$ everywhere. To make things right, one could take essentially the above approach, but start by snipping out a swath of width $\varepsilon > 0$, with sides $K^-$ and $K^+$, and then make a rigorous limiting case as $\varepsilon \to 0$.

The integrals along $K^-$ and $K^+$ cancel, and we have our result. □

FIGURE 6.10 Deformation

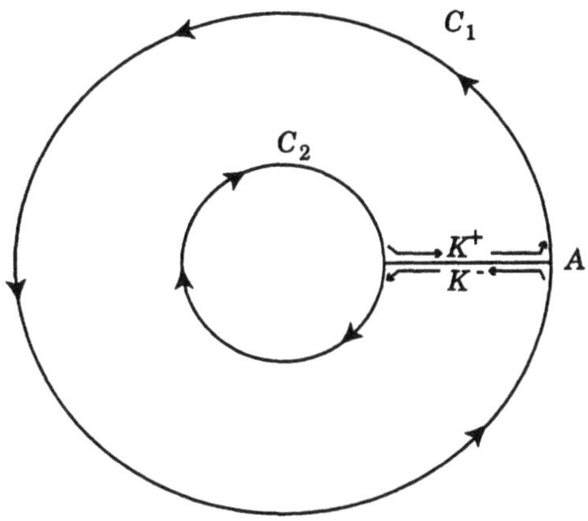


We now know that the index of a closed curve containing no equilibria is invariant under deformations, so long as equilibria are not intersected. Since the index, therefore, has little to do with the curve, what is it truly reflecting? It is really the fixed point structure inside the curve, not the curve itself, that is reflected. We make the following definition:

**Definition.** The index, $I(p)$, of a singular point, $p$, is the index of any simple closed curve surrounding $p$ that neither intersects nor encloses equilibria other than $p$.

Using an argument similar to that employed above, we shall prove:

**Theorem 9.** *The index of a closed curve equals the sum of the indices of the equilibrium points it encloses.*

**Proof.** We prove the result for two equilibria, $p_1$ and $p_2$. With $C$ the outer closed curve, surround $p_1$ and $p_2$ with curves $C_1$ and $C_2$, and make cuts $K$ and $L$, as in figure 6.11.

FIGURE 6.11 Sum of Indices

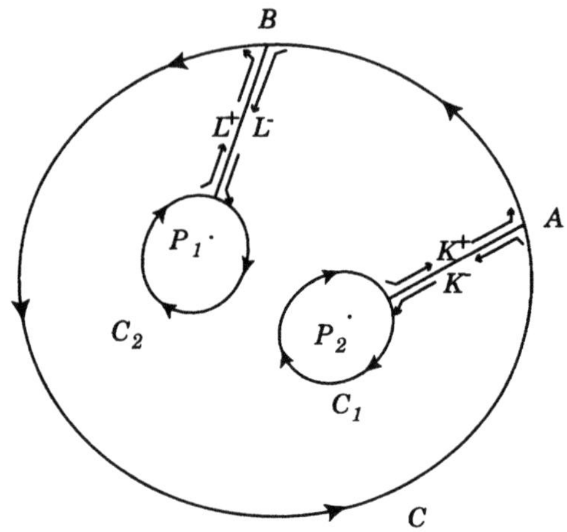


If we integrate from $A$ to $B$, down $L^-$, around $C_2$, and out $L^+$; then from $B$ to $A$, down $K^-$, around $C_1$, and out $K^+$ to the start, the closed curve described neither intersects nor encloses any equilibria, so its index is zero. Writing this out in full, but noting that $L^+$ cancels $L^-$ and that $K^+$ cancels $K^-$, we are left with

$$0 = \frac{1}{2\pi}\left(\oint_C d\Phi - \sum_{i=1}^{2}\oint_{C_i} d\Phi\right), \text{ or}$$

$$0 = I(C) - \sum_{i=1}^{2} I(p_i),$$

which quite evidently generalizes to the result we seek:

$$I(C) = \sum_{i=1}^{n} I(p_i)$$

where the $p_i$'s are singularities enclosed by $C$. □

With one more theorem we will be able to extract some very unexpected results.

**Theorem 10.** *The index of a closed periodic orbit is 1.*

**Proof.** We merely formalize the plausibility argument made earlier. Here, $C$ is a closed periodic orbit (as shown in figure 6.9 above). At every point of $C$, therefore, the field vector is precisely tangent to $C$. In one circuit around $C$, the total change in the tangent vector's—and hence the field vector's—angle of elevation, $\Phi$, to the $x$-axis is $2\pi$ radians, or

$$\oint_C d\Phi = 2\pi .$$

But this line integral is, by definition, $2\pi I(C)$. □

Theorems 8 and 10 imply:

**Corollary 2.** *A closed periodic trajectory must surround an equilibrium.*

**Proof.** Posit, to the contrary, a closed periodic orbit $C$ enclosing no equilibria. Since closed, its index must be zero, by Theorem 8. But, since $C$ is a periodic trajectory, its index is 1, by Theorem 10, yielding a contradiction. □

More general is the following "conservation law."

**Corollary 3.** *Suppose a closed periodic trajectory surrounds $C$ centers, $N$ nodes, $F$ foci, and $S$ saddles. Then,*

$$C + N + F - S = 1 . \tag{6.37}$$

**Proof.** By Theorem 9, the index of the surrounding orbit equals the sum of the indices of interior fixed points; and as observed above, every center, node, and focus contributes $+1$ to the sum; every saddle contributes $-1$, explaining the left-hand side. But, by Theorem 10, the surrounding orbit's index is 1. □

So, for example, one will not encounter a limit cycle with exactly one saddle and one focus inside, or a center and a node, as these arrangements violate (6.37). Interpreted slightly differently, (6.37) offers a second negative criterion for limit cycles. If a region is populated by centers, nodes, foci, and saddles and (6.37) is violated, then there cannot possibly be a surrounding periodic trajectory. Or, assuming $C$ to be a periodic orbit, is the phase portrait shown in figure 6.12 impossible?

FIGURE 6.12 Impossible Phase Portrait?

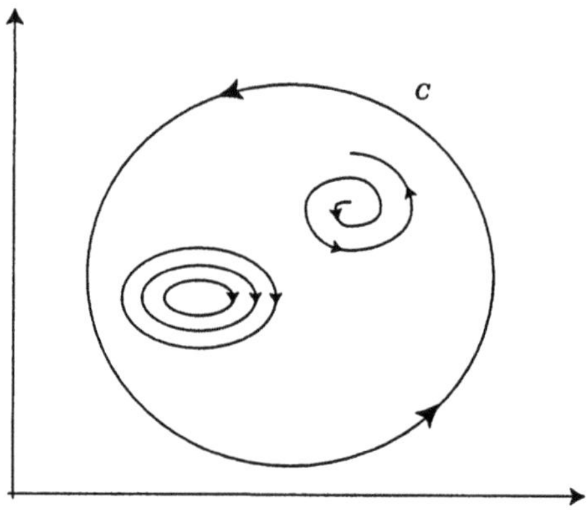


Another very unexpected result, with applications in mathematical economics,[149] is:

**Corollary 4.** *A closed periodic trajectory must enclose an odd total number of centers, nodes, foci, and saddles.*

**Proof.** Suppose to the contrary, an even number, $2n$, $n \geq 1$. Then, in the notation of Corollary 3,

$$C + N + F + S = 2n\,.$$

But, by (6.37), we have

$$C + N + F - S = 1\,.$$

Adding, we obtain

$$2(C + N + F) = 2n + 1\,,\ \text{an obvious contradiction.} \qquad \square$$

As a final result of this sort, assume, as usual, that $\Gamma$ is a closed periodic orbit intersecting no critical points. Then, from the conservation law (6.37), the following corollary relating index theory to bifurcations follows trivially.

**Corollary 5.** *If, at a bifurcation point, a saddle is created (destroyed) inside* $\Gamma$*, then a node, focus, or center must be created (destroyed).*

To conserve "topological charge" (i.e., the sum of the indices) new singularities must arise or disappear in *pairs* having opposite indices. (The parallel to particles and antiparticles in high energy physics has been suggested). In fact, the corollary

[149] See Mas-Colell (1985, p. 190).

goes through if $\Gamma$ is a simple closed curve, not necessarily an orbit. For this, much stronger, result, see Birkhoff and Rota (1989).

While this corollary flows effortlessly from the basic conservation theorem, stop and consider how utterly unapproachable it would be without index theory.

## BROUWER'S FIXED POINT THEOREM

As demonstrated in Theorem 9, the index of a curve $I(C)$ is a topological invariant in that continuous deformations of $C$ do not affect the index so long as no equilibria are intersected. Here, we imagine $C$ as being deformed over some specific underlying vector field $V$. An equivalent, but differently powerful, perspective is to fix $C$ and continuously deform the vector field $V$, ensuring as usual that no singularities are brought into contact with $C$. Of course, we have not defined "continuous deformation of a vector field." This is a fundamental notion in topology and has a special name.

**Definition.** Let $f, g : X \to Y$ be maps.[150] Then $f$ is *homotopic* to $g$ if there exists a continuous map $H(x,t), 0 \leq t \leq 1$, such that $H(x,0) = f(x)$ and $H(x,1) = g(x)$.

With this definition,[151] it can be shown—and it seems entirely plausible—that

**Fact:** The index of a curve is invariant under continuous deformations—homotopies—of the vector field so long as no equilibria are brought into contact with the curve in the course of the deformation.

Accepting this fact, index theory provides an elegant way to prove Brouwer's celebrated fixed point theorem on the closed disk in the plane. Although he does not use quite this terminology, the proof follows Arnol'd.[152]

**Theorem 11.** *Every smooth mapping $f : \bar{D} \to \bar{D}$ of the closed disk into itself has a fixed point.*

**Proof.** We imagine the disk—whose closed boundary we denote $\partial D$—to be centered at the origin. Define now the vector field

$$V(x) = f(x) - x. \tag{6.38}$$

Clearly, the fixed points of $f$ are, identically, the equilibria of $V$. The Theorem will therefore be proved if we can establish that the index of $\partial D$ in $V$ is 1. Why?

[150] Here $X$ and $Y$ are topologiocal spaces. See Royden (1988).

[151] Some treatments require smoothness. See Guillemin and Pollack (1974).

[152] Arnol'd (1973, p. 257).

Because then, by Theorem (10), $D = \text{int}(\partial D)$ must contain an equilibrium of $V$ (else $I(\partial D)$ would be zero). But that equilibrium, as just noted, is a fixed point of $f$. So, we state:

**Claim.** The index of $\partial D$ in $V$ is 1.

Now, by the Fact above, the claim will be proved if we can provide a vector field homotopic to $V$ in which $I(\partial D) = 1$. To that end, define

$$H(x,t) = tf(x) - x, 0 \leq t \leq 1\,.$$

$H$ defines a homotopy between the fields $H(x,0) = -x$ and $H(x,1) = V(x)$, and for no $t \in [0,1]$ does $H(x,t)$ have singularities on $\partial D$. But $I(\partial D)$ in the field $-x$ is obviously 1 since $-x$ is just the field with all vectors pointing to the center of $\bar{D}$. □

To summarize the proof, $f : \bar{D} \to \bar{D}$ has a fixed point if and only if $V$ has an equilibrium (by construction of $V$). In turn, $V$ has an equilibrium in $\bar{D}$ if $I(\partial D)$ in $V$ is 1 (by Theorem 10). And $I(\partial D)$ in $V$ is 1 if $I(\partial D)$ is 1 in any field homotopic to $V$. With $H(x,t)$ defined as above, $H(x,0) = -x$ is homotopic to $V$ and $I(\partial D)$ in $-x$ is obviously 1.

Penetrating as the theory is when applied to vector fields in the plane, even more startling results emerge when we consider vector fields on more general objects, notably closed surfaces like the sphere and torus.

## A GLIMPSE BEYOND THE PLANE

In particular, I will conclude with a *very* informal presentation of the beautiful Poincaré-Hopf index theorem, which nicely connects our work on index theory to differential topology.

Perhaps the most classical place to begin is with Euler's well-known formula (which was apparently known to Descartes)[153] that for closed convex polyhedral surfaces "like" the pyramid and cube,[154] if $V$, $E$, and $F$ are the numbers of vertices, edges, and faces, then

$$V - E + F = 2\,. \tag{6.39}$$

Now, think of a sphere surrounded by a closed "approximating" polyhedral surface, all of whose faces are triangles that fit together "nicely"—so that triangles intersect only in a common vertex or an entire common edge. By (6.39), if you count vertices, subtract edges, and add faces, you will again find that the sum is 2. This number, of

[153] See Fréchet and Fan (1967, p. 21).

[154] In topology, we do not use the word "like" lightly. For a punctilious characterization of exactly those polyhedra to which the result applies, see Armstrong (1983).

course, does not change if we allow "rubber triangles"—which we imagine pressing down to exactly cover the sphere—and count curved faces and edges. This is the idea of a *triangulation* of a surface. You could also imagine etching curved triangles all over the spherical surface in the "nice" way I mentioned above, as shown in figure 6.13.

FIGURE 6.13 The Triangulations of the Sphere.

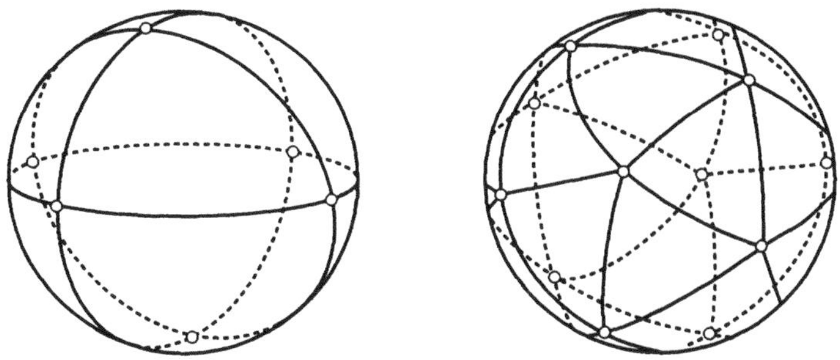

Source: Based on Shashkin (1991, p. 55).

To be precise, however, a triangulation of a (compact) surface $S$ is a covering of $S$ by a finite family of closed sets $\{T_i\}$, each of which is the homeomorphic *image* of a triangle in the plane.[155] And, again, we require that two distinct "triangles" (images) be disjoint or have in common either a single "vertex" or one entire "edge," where these are also understood to be image objects. Leaving out the quotes, on this understanding, we can count $V-E+F$ just as before. It is a deep fact that for a given surface, $S$, the number thus obtained is independent of the triangulation;[156] it is a characteristic of the surface itself—the so-called Euler characteristic $\chi(S)$. The $n$-dimensional definition is

$$\chi(S) = \sum_{i=0}^{n}(-1)^i \alpha_i \,, \tag{6.40}$$

where $\alpha_i$ is the number of "faces of dimension $i$."[157] For $n = 2$, $\alpha_0$, $\alpha_1$, and $\alpha_2$ are, respectively, the numbers of vertices (dimension 0), edges (dimension 1), and

[155] Massey (1967).

[156] It is, in fact, a theorem (Rado, 1925) that a triangulation is possible.

[157] $i$-simplexes.

faces (dimension 2) in the triangulation. So in two dimensions, we recover Euler's alternating sum:

$$\chi(S) = V - E + F\,.$$

And, the generalized Euler theorem is that for all surfaces homeomorphic to the sphere—such as the cube and pyramid above—the Euler characteristic is 2.

Naturally, we will be interested in the Euler characteristics of other smooth closed surfaces like the torus. And we expect the Euler characteristic, being topological, to depend on the connectedness of the surface, a property captured in the so-called *genus*.[158]

We met the notion of connectedness in proving that the index of a closed curve equals the sum of the indices of enclosed singularities—we used a line segment to "slice" a multiply connected domain, creating a simply connected one to which a more basic theorem applied. But, the *number of slices* needed was clearly a topological invariant having to do with the connectedness of the domain.

Similarly, for closed two-manifolds (which we will define shortly) like the sphere, torus, and so on, one can ask: What is the maximum number of, now, simple closed *curves* one can draw on the surface *without* dissecting it into disconnected parts? This number is the *genus* of the manifold. For the sphere, it is clearly zero, since even one simple closed curve will dissect it. Two simple closed curves are needed to dissect the torus, so its genus is 1, and so on, as shown in figure 6.14.

FIGURE 6.14 Classification of Oriented Closed Two-Manifolds

Genus 0 (homeomorphic to a sphere with no handles)

Genus 1 (homeomorphic to a sphere with one handle)

Genus 2 (homeomorphic to a sphere with two handles)

Genus 3 (homeomorphic to a sphere with three handles)

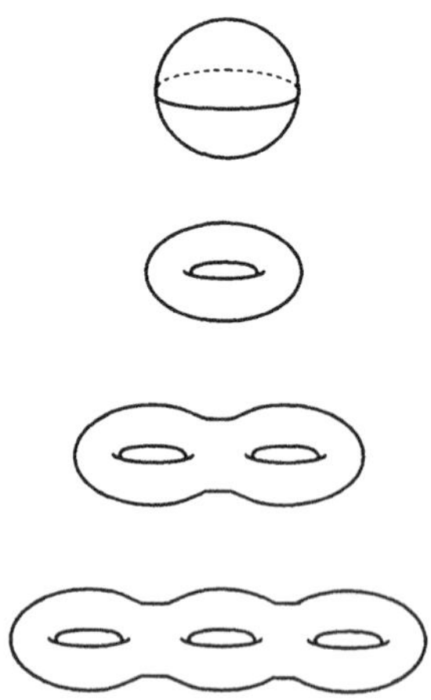

Source: Based on Guillemin and Pollack (1974, p. 124).

[158]The Euler characteristic also depends on the *orientability* of the surface, a topic into which we will not enter here. All surfaces are assumed to be oriented, even though I will often say so explicitly just to drive home the point that it matters. The classic nonorientable surface is, of course, the Moebius strip.

The figure reflects a fundamental classification theorem in topology: *every compact oriented boundaryless two-manifold is homeomorphic to one of these surfaces—a sphere with $n \geq 0$ handles.*[159] We know from above that the Euler characteristic of the sphere is 2. Its genus is zero. The general relationship is:

**Theorem 12.** *Oriented closed surfaces*[160] *of genus $g$ have Euler characteristic $2 - 2g$.*[161]

We seem to have wandered far from index theory, which concerned singularities of vector fields. To proceed further and connect all of this to index theory, we need to be able to define vector fields on surfaces of the sort we have discussed. And we are going to want to do calculus on them, which puts us in the world of differential topology. Obviously, we all learned calculus on the Euclidean plane. And we will be able to do calculus on objects that are "locally Euclidean" which is certainly how practitioners think of *manifolds*: surfaces that are, locally, smooth deformations of the plane. In fact, all the closed surfaces we have been discussing are two-manifolds. Technically, a two-manifold is a connected Hausdorff space,[162] $M$, each point of which has a neighborhood homeomorphic to an open set in $\mathcal{R}^2$. This is what one means by "locally Euclidean." And, in turn, a vector field on $M$ is simply a smooth ($C^1$) assignment of a tangent vector to each point $x \in M$, just as in vector calculus. But, it will pay to be a bit more painstaking. Following Smale (1969), we associate to each point $x \in M$ a two-dimensional vector space, $T_x(M)$, the so-called tangent space of $M$ at $x$. For a two-manifold, like the sphere, $T_x(M)$ is the plane tangent to $M$ at $x$. A vector field $V(x)$ on $M$ is a $C^1$ assignment to each $x \in M$ of a tangent vector—that is, a vector lying in $T_x(M)$.

At singularities $\bar{x}$ where $V(\bar{x}) = 0$, the field may exhibit sinks, sources, centers, saddles, and more complex behaviors—phase portraits—*on* $M$. Technically, we do not know how to compute indices *on* $M$. But, $M$ is an object that is locally homeomorphic to the Euclidean space, $\mathcal{R}^2$, where we do know how to compute indices. So, we imagine doing the same thing on a small surface element of the tangent space $T_x(M)$ at $\bar{x}$.[163] As Guillemin and Pollack put it, "Looking at the manifold locally,

[159] Indeed, for two-manifolds, diffeomorphic.

[160] Though, again, we will not delve into it, the general relationship between genus $g$, Euler characteristic $\chi$, and orientability is

$$\chi(S) = \begin{cases} 2 - 2g & \text{if } S \text{ is orientable ,} \\ 2 - g & \text{if } S \text{ is nonorientable .} \end{cases}$$

[161] See Courant and Robbins (1963) for a very nice intuitive development.

[162] A topological space is Hausdorff if it satisfies the following condition: Given two distinct points $x$ and $y$, there exist open sets $O_1$ and $O_2$ such that $x \in O_1$ and $y \in O_2$ and $O_1 \cap O_2 = \phi$. See Royden (1988, p. 178).

[163] This, again, is standard procedure in surface integration, for example.

we see essentially a piece of Euclidean space, so we simply read off the index as if the vector field were Euclidean."[164]

With all this in place, we can state the Poincaré-Hopf index theorem. As you might imagine, there are many formulations; we follow Aleksandrov's.[165]

**Theorem 13.** (Poincaré-Hopf). *If, on a given oriented two-manifold, a continuous vector field is defined having only a finite number of singular points, then the sum of their indices equals the Euler characteristic of the manifold.*

This is remarkable because the Euler characteristic is a topological property of the manifold only, and would appear to have absolutely nothing to do with flows, or vector fields, *on* the manifold. How does the manifold, $M$, already "know" so much about the singularity structure of vector fields definable on $M$? Clearly, deep things are afoot.

One immediate consequence of the theorem is that vector fields having no singularities are possible only on manifolds with Euler characteristic zero—or genus 1; namely, the torus (and, in fact, the Klein bottle, a nonorientable "one-sided" torus). Notably, it is impossible to construct a nonvanishing field on the sphere. This result is sometimes affectionately called "the hairy ball theorem," the interpretation being that any attempt to comb smooth a "hairy ball" must leave at least one bald spot. Baldness can be avoided on the torus, as shown in figure 6.15.

FIGURE 6.15 Hairy Ball Theorem.

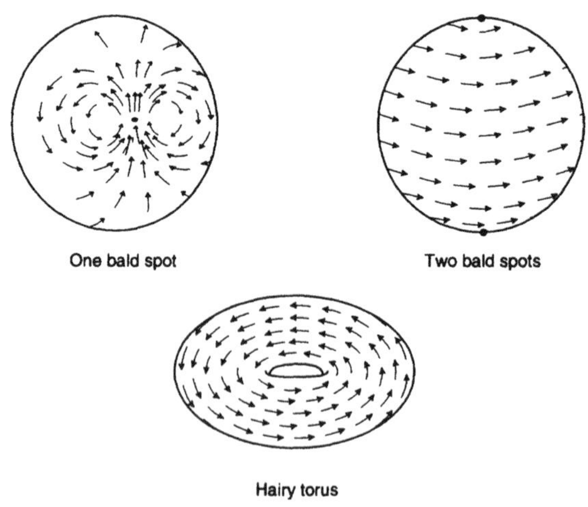


Source: Based on Armstrong (1983, p. 198).

[164] Guillemin and Pollack (1974, p. 133).

[165] See Aleksandrov, et al. (1963, p. 216).

Another interpretation on the sphere is that "somewhere on the surface of the earth the wind isn't blowing." For an interesting application of the theorem to chemical and ecological networks, see Glass (1975).

We certainly cannot prove the Poincaré-Hopf index theorem here. But, if you will grant that there is something generic about a certain map, the so-called Lefschetz map, then at least a very suggestive example will be in hand. Following Guillemin and Pollack (1974), consider the genus 4 two-manifold in figure 6.16.

FIGURE 6.16 A Two-Manifold of Genus 4.

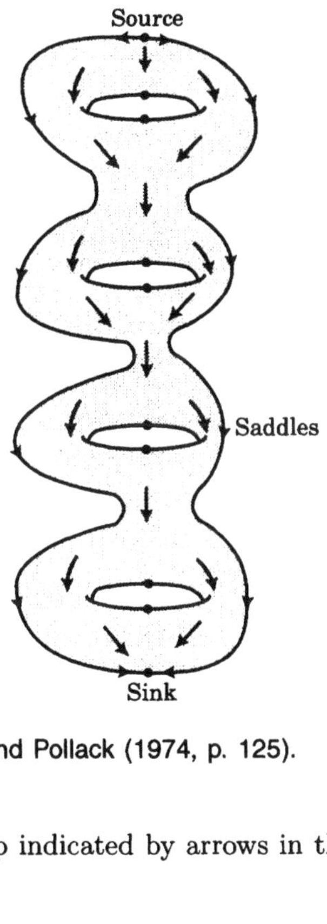


Source: Based on Guillemin and Pollack (1974, p. 125).

Their description of the map indicated by arrows in the figure is clear, and appetizing!

> "We 'construct' a Lefschetz map on the surface of genus $k$ as follows. Stand the surface vertically on one end, and coat it evenly with hot fudge topping. Let $f_t(x)$ denote the oozing trajectory of the point $x$ of fudge as time $t$ passes. At time 0, $f_0$ is just the identity. At time $t > 0$, $f_t$ is a Lefschetz

> map with one source at the top, one sink at the bottom, and saddlepoints at the top and bottom of each hole" (Guillemin and Pollack, 1974, p. 125).

As I said, this oriented manifold is obviously of genus 4. Notice that in this case, the genus (the number of handles) is also the number of holes. Now, let us sum the indices.

By our previous work (and the crucial fact that a manifold is locally Euclidean), we can directly count plus two for the sink and the source, and minus one for each saddle; that is, minus two for each "hole" of which there are, in general, $h$. So, the global sum of the indices is $2 - 2h$. But, the number of holes is the genus of the manifold. So, by Theorem 12, $2 - 2h$ is also the Euler characteristic, as predicted by Poincaré-Hopf!

## A CONCLUDING PERSPECTIVE

Many of the topics treated in this lecture are far more advanced than the mathematics applied in other lectures. While certain of the topics may seem highly abstract, the history of science shows that applications of pure mathematics are hard to anticipate. Who imagined that complex numbers, non-Euclidean geometries, and infinite-dimensional spaces would find powerful applications in physics?[166] Against that background, it would be naive to preclude the applicability of even the most abstract mathematics to social science. And meanwhile, of course, the mathematics is its own reward.

[166]On this, see Eugene Wigner's essay, "The Unreasonable Effectiveness of Mathematics in the Natural Sciences." (Wigner, 1963).

## REFERENCES


Aleksandrov, A. D., A. N. Kolmogorov, and M. A. Lavrent'ev. 1989. *Mathematics: Its Content, Methods, and Meaning.* Translated by S. H. Gould, K. A. Hirsch, and T. Bartha. Cambridge, MA: MIT Press.

Anderson, Roy M., and Robert M. May. 1991. *Infectious Diseases of Humans: Dynamics and Control.* New York: Oxford University Press.

Armstrong, M. A. 1983. *Basic Topology.* New York: Springer-Verlag.

Arnol'd, Vladimir I. 1973. *Ordinary Differential Equations.* Translated by Richard A. Silverman. Cambridge, MA: MIT Press.

Arnol'd, Vladimir I. 1991. *Ordinary Differential Equations.* Translated by Roger Cooke. New York: Springer-Verlag.

Arrowsmith, D. K., and C. M. Place. 1990. *An Introduction to Dynamical Systems.* New York: Cambridge University Press.

Axelrod, Robert, 1984. *The Evolution of Cooperation.* New York: Basic Books.

Axelrod, Robert. 1986. "An Evolutionary Approach to Norms." *Amer. Poli. Sci. Rev.* **80** (December): 1095–1111.

Axelrod, Robert. 1987. "The Evolution of Strategies in the Iterated Prisoner's Dilemma." In *Genetic Algorithms and Simulated Annealing*, edited by L. Davis. London: Pitman Publishing.

Bailey, Norman T. J. 1975. *The Mathematical Theory of Infectious Diseases and Its Applications.* New York: Hafner Press.

Bailey, Norman T. J. 1957. *The Mathematical Theory of Epidemics.* New York: Hafner Publishing.

Beatty, Jack. 1986. "Along the Western Front." *Atlantic Monthly* **258**: 112–115.

Beltrami, Edward. 1987. *Mathematics for Dynamic Modeling.* San Diego, CA: Academic Press.

Birkhoff, Garrett, and Gian-Carlo Rota. 1989. *Ordinary Differential Equations*, 4th ed. New York: John Wiley & Sons.

Borrelli, Robert L., and Courtney S. Coleman. 1987. *Differential Equations: A Modeling Approach.* Englewood Cliffs, NJ: Prentice-Hall.

Braun, M. 1983. *Differential Equations and Their Applications: An Introduction to Applied Mathematics*, 3rd ed. New York: Springer-Verlag.

Britton, N. F. 1986. *Reaction-Diffusion Equations and Their Applications to Biology.* New York: Academic Press.

Cavalli-Sforza, L. L., and M. W. Feldman. 1981. *Cultural Transmission and Evolution: A Quantitative Approach.* Princeton, NJ: Princeton University Press.

Clark, Colin W. 1990. *Mathematical Bioeconomics: The Optimal Management of Renewable Resources*, 2nd ed. New York: John Wiley & Sons.

Coleman, Courtney S. 1978. "Hilbert's 16th Problem: How Many Cycles?" In *Differential Equation Models*, edited by Martin Braun et al., 279–297. New York: Springer-Verlag.

Courant, Richard, and Herbert Robbins. 1963. *What is Mathematics?* New York: Oxford University Press.
Devaney, Robert L. 1989. *An Introduction to Chaotic Dynamical Systems*, 2nd ed. Reading, MA: Addison-Wesley.
Dietz, Klaus. 1967. "Epidemics and Rumours: A Survey." *J. Roy. Stat. Soc., Ser. A* **130**: 505–528.
Downs, George W., ed. 1994. *Collective Security Beyond the Cold War*. Ann Arbor: University of Michigan Press.
Downs, George W., and David M. Rocke. 1990. *Tacit Bargaining, Arms Races, and Arms Control.* Ann Arbor: University of Michigan Press.
Edelstein-Keshet, Leah. 1988. *Mathematical Models in Biology*. Berkhäuser Mathematical Series. New York: Random House.
Epstein, Joshua M. 1985. *The Calculus of Conventional War: Dynamic Analysis Without Lanchester Theory.* Washington, DC: Brookings Institution.
Epstein, Joshua M. 1990. *Conventional Force Reductions: A Dynamical Assessment.* Washington, DC: Brookings Institution.
Epstein, Joshua M. 1993. "The Adaptive Dynamic Model of Combat." In *1992 Lectures in Complex Systems,* edited by L. Nadel and D. Stein, 437–459. Santa Fe Institute Studies in the Sciences of Complexity, Vol. 5. Reading, MA: Addison-Wesley.
Epstein, Joshua M., and Robert Axtell. 1996. *Growing Artificial Societies: Social Science from the Bottom Up*. Washington, DC: Brookings Institution and Cambridge, MA: MIT Press.
Fife, Paul C. 1979. *Mathematical Aspects of Reacting and Diffusing Systems.* New York: Springer-Verlag.
Forrest, Stephanie, and Gottfried Mayer-Kress. 1991. "Genetic Algorithms, Nonlinear Dynamical Systems, and Models of International Security." In *Handbook of Genetic Algorithms*, edited by Lawrence Davis, 166–185. New York: Van Nostrand.
Fréchet, Maurice, and Ky Fan. 1967. *Initiation to Combinatorial Topology.* Translated by Howard W. Eves. Boston: Prindle, Weber and Schmidt.
Freeman, James A., and David M. Skapura. 1991. *Neural Networks: Algorithms, Applications, and Programming Techniques.* Reading, MA: Addison-Wesley.
Gamelin, Theodore W., and Robert Everest Greene. 1983. *Introduction to Topology.* The Saunders Series. New York: CBS College Publishing.
Gause, G. F. 1934. *The Struggle for Existence.* Baltimore: Williams & Wilkins.
Gell-Mann, Murray. 1988. "The Concept of the Institute." In *Emerging Syntheses in Science*, edited by David Pines. Santa Fe Institute Studies in the Sciences of Complexity, Vol. 1, 4. Reading, MA: Addison-Wesley.
Glass, L. 1975. "A Topological Theorem for Nonlinear Dynamics in Chemical and Ecological Networks." *Proc. Nat. Acad. Sci. USA* **72**: 2856–57
Goh, B. S. 1979. "Stability in Models of Mutualism." *Am. Natur.* **113(2)**: 261–275.

Goodwin, R. M. 1967. "A Growth Cycle." In *Socialism, Capitalism, and Economic Growth*, edited by C. H. Feinstein, 54–58. Cambridge: Cambridge University Press.

Gould, James L., and Carol Grant Gould. 1989. *Sexual Selection*. New York: Scientific American Library.

Groetsch, Charles W. 1980. *Elements of Applicable Functional Analysis*. New York: Marcel Dekker.

Guckenheimer, John, and Philip Holmes. 1990. *Nonlinear Oscillations, Dynamical Systems, and Bifurcations of Vector Fields*. New York: Springer-Verlag, 1983. Corrected 3rd printing.

Guillemin, Victor, and Alan Pollack. 1974. *Differential Topology*. Englewood Cliffs, NJ: Prentice-Hall.

Hale, Jack K., and Hüseyin Koçak. 1991. *Dynamics and Bifurcations*. New York: Springer-Verlag.

Hethcote, Herbert W. 1976. "Qualitative Analyses of Communicable Disease Models." *Math. Biosci.* **28**: 335–356.

Hethcote, Herbert W. 1989. "Three Basic Epidemic Models." In *Applied Mathematical Ecology*, edited by Simon A. Levin, Thomas G. Hallam, and Louis J. Gross, 119–144. New York: Springer-Verlag.

Hethcote, Herbert W., and Simon A. Levin. 1989. "Periodicity in Epidemiological Models." In *Applied Mathematical Ecology*, edited by S. A. Levin, T. G. Hallam, and L. J. Gross, 193–211. New York: Springer-Verlag.

Hethcote, Herbert W., and James A. Yorke. 1980. *Gonorrhea Transmission Dynamics and Control*. New York: Springer-Verlag.

Hildebrand, Francis B. 1976. *Advanced Calculus for Applications*, 2nd ed. Englewood Cliffs, NJ: Prentice-Hall.

Hirsch, Morris W. 1984. "The Dynamical Systems Approach to Differential Equations." *Bull. Am. Math. Soc.* **2(1)** (July): 1–64.

Hirsch, Morris W., and Stephen Smale. 1974. *Differential Equations, Dynamical Systems, and Linear Algebra*. San Diego, CA: Academic Press.

Hofbauer, Josef, and Karl Sigmund. 1988. *The Theory of Evolution and Dynamical Systems*. New York: Cambridge University Press.

Hoover, Dean, and David Kowaleski. 1992. "Dynamic Models of Dissent and Repression." *J. Conflict Resolution* **36(1)** (March): 150–192.

Huang, Xun-Cheng, and Stephen J. Merrill. 1989. "Conditions for Uniqueness of Limit Cycles in General Predator-Prey Systems." *Math. Biosci.* **96** : 47–60.

Huntley, Ian D., and R. M. Johnson. 1983. *Linear and Nonlinear Differential Equations*. Chichester, West Sussex: Ellis Horwood Limited.

Il'yashenko, Yu. S. 1991. *Finiteness Theorems for Limit Cycles*. Translations of Mathematical Monographs, Vol. 94. Providence, R.I.: American Mathematical Society.

Jackson, E. Atlee. 1989. *Perspectives of Nonlinear Dynamics*, Vols. 1 & 2. New York: Cambridge University Press.

Jordan, D. W., and P. Smith. 1987. *Nonlinear Ordinary Differential Equations*, 2nd ed. New York: Oxford University Press.

Kaufmann, William W. 1983. "The Arithmetic of Force Planning." In *Alliance Security: NATO and the No-First-Use Question*, edited by J. D. Steinbruner and L. V. Sigal. Washington, DC: Brookings Institution.

Kermack, W. O., and A. G. McKendrick. 1927. "Contributions to the Mathematical Theory of Epidemics." *Proc. Roy. Stat. Soc., Ser. A* **115**: 700–721.

Krasner, Stephen. 1983. *International Regimes*. Ithaca, NY: Cornell University Press.

Kreyszig, Erwin. 1978. *Introductory Functional Analysis with Applications*. New York: John Wiley & Sons. Reprinted in 1989.

Kupchan, Charles A., and Clifford A. Kupchan. 1995. "The Promise of Collective Security." *Internatl. Security* **20(1)** (Summer): 52–70.

Kupchan, Charles A., and Clifford A. Kupchan. 1991. "Concerts, Collective Security, and the Future of Europe." *Internatl. Security* **16(1)** (Summer): 114–161.

Lanchester, F. W. 1916. *Aircraft in Warfare: The Dawn of the Fourth Arm*. London: Constable.

Lanchester, F. W. 1956. "Mathematics in Warfare." In *The World of Mathematics*, edited by James R. Newman, Vol. 4, 2136–2137. New York: Simon & Schuster.

Lefschetz, Solomon. 1977. *Differential Equations: Geometric Theory*. New York: Dover Publications, Inc.

Lorenz, Hans Walter. 1989. *Nonlinear Dynamical Economics and Chaotic Motion*. Berlin: Springer-Verlag.

Mansfield, Edwin. 1961. "Technical Changes and the Rate of Imitation." *Econometrica* **29(4)** (October): 741–766.

Marsden, Jerrold E. 1974. *Elementary Classical Analysis*. New York: W.H. Freeman and Company.

Marsden, Jerrold E., and M. McCracken. 1976. *The Hopf Bifurcation and Its Applications*. New York: Springer-Verlag.

Marsden, Jerrold E., and Anthony J. Tromba. 1976. *Vector Calculus*. San Francisco: W. H. Freeman.

Mas-Colell, Andreu. 1985. *The Theory of General Economic Equilibrium*. New York: Cambridge University Press.

Massey, W. S. 1967. *Algebraic Topology: An Introduction*. New York: Springer-Verlag.

May, Robert M. 1974. *Stability and Complexity in Model Ecosystems*, 2nd ed. Princeton, NJ: Princeton University Press.

May, Robert M. 1981. "Models for Two Interacting Populations." In *Theoretical Ecology*, edited by Robert M. May. London: Blackwell Scientific Publications.

May, R. M. 1983. "Parasitic Infections as Regulators of Animal Populations." *Amer. Sci.* **71**: 36–45.

Mayer-Kress, Gottfried. 1992. "Nonlinear Dynamics and Chaos in Arms Race Models." In *Modeling Complex Phenomena*, edited by Lui Lam and Vladimir Naroditsky. New York: Springer-Verlag.

Maynard Smith, John. 1982. *Evolution and the Theory of Games*. New York: Cambridge University Press. Reprinted in 1986, 1989.
Maynard Smith, John. 1989. *Evolutionary Genetics*. Oxford University Press.
McNeill, W. H. 1976. *Plagues and Peoples*. New York: Anchor Press/Doubleday.
Mearsheimer, John J. 1994. "The False Promise of International Institutions." *Internatl. Security* **19(3)** (Winter): 5–49.
Minorsky, Nicholas. 1962. *Nonlinear Oscillations*. Princeton, NJ: D. Van Nostrand.
Morel, Benoit. 1991. "Modelling U.S.-Soviet Relations." Draft Analysis. Carnegie-Mellon University.
Murray, J. D. 1989. *Mathematical Biology*. New York: Springer-Verlag.
Naylor, Arch W., and George R. Sell. 1982. *Linear Operator Theory in Engineering and Science*. New York: Springer-Verlag.
Niou, Emerson M. S., and Peter C. Ordeshook. 1991. "Realism versus Neoliberalism: A Formulation." *Amer. J. Poli. Sci.* **35(2)** (May): 481–511.
Olsen, L. F., and W. M. Schaffer. 1990. "Chaos versus Noisy Periodicity: Alternative Hypotheses for Childhood Epidemics." *Science* (August 3, 1990): 499–504.
Olson, Harry F. 1958. *Dynamical Analogies*, 2nd ed. Princeton, NJ: D. Van Nostrand.
Olson, Mancur. 1965. *The Logic of Collective Action*. Cambridge, MA: Harvard University Press.
Osipov, M. 1915. "The Influence of the Numerical Strength of Engaged Forces on Their Casualties." Originally published in the Tzarist Russian journal *Military Collection* (June-October). Also in translation as CAA-RP-91-2, translated by Robert L. Helmbold and Allan S. Rehm, U.S. Army Concepts Analysis Agency, 1991.
Poincaré, Henri. 1881, 1882, 1885, 1886. "Mémoire sur les courbes définie par une équation différentielle," I, II, III, and IV. *J. Math. Pures Appl.* **(3)7** (1881) pp. 375–422; **(3)8** (1882) pp. 251–86; **(4)1** (1885) pp. 167–244; **(4)2** (1886) pp. 151–217 (VII 0, 1, 3, 4, 5, 6-7, 13-14; VIII 3; IX 0).
Rado, T. 1925. "Über den Begriff der Riemannschen Fläche." *Acta Litt. Sci. Szeged.* **2**: 101–121.
Rappaport, Anatol. 1974. *Fights, Games, and Debates*. Ann Arbor: University of Michigan Press.
Rashevsky, N. 1947. *Mathematical Theory of Human Relations: An Approach to Mathematical Biology of Social Phenomena*. Bloomington, IN: The Principia Press.
Rashevsky, Nicolas. 1951. *Mathematical Biology of Social Behavior*. Chicago: University of Chicago Press.
Richardson, Lewis F. 1939. *Generalized Foreign Politics*. Cambridge: The University Press.
Richardson, Lewis F. 1960. *Arms and Insecurity: A Mathematical Study of the Causes and Origins of War*. Pittsburgh: Boxwood Press.
Robinson, Michael H. 1992. "An Ancient Arms Race Shows No Sign of Letting Up." *Smithsonian* **23(1)**: 74–82.

Rosen, Robert. 1970. *Dynamical System Theory in Biology, Vol. I: Stability Theory and Its Applications.* New York: Wiley Interscience.

Roughgarden, Johnathan. 1979. *Theory of Population Genetics and Evolutionary Ecology: An Introduction.* New York: Macmillan.

Royden, H. L. 1988. *Real Analysis*, 3rd ed. New York: Macmillan.

Rumelhart, David E., and James L. McClelland. 1986. *Parallel Distributed Processing*, Vol. 1. Cambridge, MA: The MIT Press.

Samuelson, Paul A. 1971. "Generalized Predator-Prey Oscillations in Ecological and Economic Equilibrium." *Proc. Natl. Acad. Sci. USA* **68(5)** (May): 980–981. Also in *The Collected Scientific Papers of Paul A. Samuelson*, edited by Robert C. Merton, Vol. III, 487–490. Cambridge, MA: MIT Press.

Samuelson, Paul A. 1972. "Maximum Principles in Analytical Economics." In *The Collected Scientific Papers of Paul A. Samuelson*, edited by Robert C. Merton, Vol. III, 8–9. Nobel Memorial Lecture, Dec. 11, 1970. Cambridge, MA: MIT Press.

Shashkin, Yu. A. 1991. "Fixed Points." Translated by Viktor Minachin. *Mathematical World* **2**.

Smale, Steve. 1980. "What is Global Analysis?" In *The Mathematics of Time.* New York: Springer-Verlag.

Smoller, Joel. 1983. *Shock Waves and Reaction–Diffusion Equations.* New York: Springer-Verlag.

Stares, Paul B. 1996. *Global Habit: The Drug Problem in a Borderless World.* Washington, DC: Brookings Institution.

Tamayo, Pablo, and Hyman Hartman. 1989. "Cellular Automata, Reaction-Diffusion Systems and the Origin of Life." In *Artificial Life,* edited by Christopher G. Langton. Santa Fe Institute Studies in the Sciences of Complexity, Proc. Vol. VI, 105–124. Reading, MA: Addison-Wesley.

Verhulst, Ferdinand. 1990. *Nonlinear Differential Equations and Dynamical Systems.* Berlin: Springer-Verlag.

Waltman, Paul. 1974. *Deterministic Threshold Models in the Theory of Epidemics.* Lecture Notes in Biomathematics, Vol. 1. New York: Springer-Verlag.

Waltman, Paul. 1986. *A Second Course in Elementary Differential Equations.* Orlando, FL: Academic Press/Harcourt Brace Jovanovich.

Weiss, Herbert K. 1966. "Combat Models and Historical Data: The U.S. Civil War." *Oper. Res.* **14**: 788.

Wiggins, S. 1990. *Introduction to Applied Nonlinear Dynamical Systems and Chaos.* New York: Springer-Verlag.

Wigner, E. 1960. "The Unreasonable Effectiveness of Mathematics in the Natural Sciences." *Commun. Pure Appl. Math.* **13**: 1–14.

Willard, D. 1962. "Lanchester as Force in History: An Analysis of Land Battles of the Years 1618–1905." Technical Paper RAC-TP-74, Research Analysis Corp., Bethesda, MD.

Wilson, Edward O. 1975. *Sociobiology.* Cambridge, MA: Harvard University Press.

Wilson, Edward O. 1978. *On Human Nature*, chap. 7. Cambridge, MA: Harvard University Press.

Wolfram, Stephen. 1991. *Mathematica: A System for Doing Mathematics by Computer*, 2nd ed. Reading, MA: Addison-Wesley.

Wrangham, Richard W. 1988. "War in Evolutionary Perspective." In *Emerging Syntheses in Science*, edited by David Pines. Santa Fe Institute Studies in the Sciences of Complexity, Proc. Vol. I. Reading, MA; Addison-Wesley.

# Index

F



G



H



I



J



K



L



M

For Product Safety Concerns and Information please contact our EU representative GPSR@taylorandfrancis.com
Taylor & Francis Verlag GmbH, Kaufingerstraße 24, 80331 München, Germany

www.ingramcontent.com/pod-product-compliance
Ingram Content Group UK Ltd.
Pitfield, Milton Keynes, MK11 3LW, UK
UKHW022201190926
475702UK00003B/144
* 9 7 8 0 3 6 7 3 2 0 0 1 0 *